EPISTEMIC LOGIC

Other interview books from Automatic Press ♦ VIP

Formal Philosophy
edited by Vincent F. Hendricks & John Symons
November 2005

Masses of Formal Philosophy
edited by Vincent F. Hendricks & John Symons
October 2006

Political Questions: 5 Questions for Political Philosophers
edited by Morten Ebbe Juul Nielsen
December 2006

Philosophy of Technology: 5 Questions
edited by Jan-Kyrre Berg Olsen & Evan Selinger
February 2007

Game Theory: 5 Questions
edited by Vincent F. Hendricks & Pelle Guldborg Hansen
April 2007

Philosophy of Mathematics: 5 Questions
edited by Vincent F. Hendricks & Hannes Leitgeb
January 2008

Philosophy of Computing and Information: 5 Questions
edited by Luciano Floridi
Sepetmber 2008

Philosophy of the Social Sciences: 5 Questions
edited by Diego Ríos & Christoph Schmidt-Petri
September 2008

Epistemology: 5 Questions
edited by Vincent F. Hendricks & Duncan Pritchard
September 2008

Complexity: 5 Questions
Carlos Gershenson
November 2008

Probability and Statistics: 5 Questions
edited by Alan Hajék and Vincent F. Hendricks
September 2009

See all published and forthcoming books in the 5 Questions series at
www.vince-inc.com/automatic.html

EPISTEMIC LOGIC
5 QUESTIONS

edited by

Vincent F. Hendricks

Olivier Roy

Automatic Press ♦ $\frac{V}{I}$P

Automatic Press ♦ $\frac{V}{I}$P

Information on this title: www.vince-inc.com/automatic.html

© Automatic Press / VIP 2010

This publication is in copyright. Subject to statuary exception and to the provisions of relevant collective licensing agreements, no reproduction of any part may take place without the written permission of the publisher.

First published 2010

Printed in the United States of America and the United Kingdom

ISBN-10 87-92130-24-0 paperback
ISBN-13 978-87-92130-24-2 paperback

The publisher has no responsibilities for the persistence or accuracy of URLs for external or third party Internet Web sites referred to in this publication and does not guarantee that any content on such Web sites is, or will remain, accurate or appropriate.

Typeset in $\LaTeX 2_\varepsilon$
Graphic design by Olivier Roy & Vincent F. Hendricks

Contents

Preface   iii

Acknowledgements   v

1 Horacio Arló-Costa   1

2 Sergei Artemov   11

3 Robert Aumann   21

4 Johan van Benthem   35

5 Giacomo Bonanno   47

6 Adam Brandenburger   59

7 Hans van Ditmarsch   73

8 Melvin Fitting   83

9 Paul Gochet   93

10 Joseph Y. Halpern   111

11 Sven Ove Hansson   123

12 Aviad Heifetz   133

13 Jaakko Hintikka   145

14 Wiebe van der Hoek   149

15 Wolfgang Lenzen   153

16 John-Jules Ch. Meyer   157

17 Lawrence S. Moss   165

18 Rohit Parikh	173
19 Wlodek Rabinowicz	181
20 R. Ramanujam	191
21 Krister Segerberg	201
22 Yoav Shoham	223
23 John Sowa	231
24 Robert Stalnaker	243
25 Timothy Williamson	249
About the Editors	263

Preface

This volume started with a phone call. In April 2009 I (Olivier Roy) was visiting Roskilde University to give a seminar on Epistemic Logic and the Foundations of Game Theory. During one of the sessions, the phone of one of Vincent's (then) PhD students rang. It was Vincent, who wanted to talk to me "N-O-W". A bit puzzled by this sudden and rather urgent interruption, I excused myself, went outside the lecture room, and took the phone:

Olivier: Hi.

Vincent: Olivier, this is Vincent. *I have business proposition for you.* Let's do an *Epistemic Logic: 5 Questions*.

I resumed the lecture about two minutes later, with an exiting and almost fully worked out project in hand.

◆

Epistemic Logic was a natural choice for a volume in the *5 Questions* series. Since 2005 the series has covered very diverse areas of philosophy, and even forayed in economics, mathematics, computer sciences, social sciences, semiotics and biology. With the present volume it comes back to the themes dear to the first two, *Formal Philosophy* and *Masses of Formal Philosophy*: formal methods and foundational research.

The interviews gathered show that epistemic logic raises issues of interest for a broad variety of researchers: from philosophy to game theory, computer science, AI and linguistic. This is not a coincidence. On the one hand, knowledge, belief and information are key notions in understanding strategic interaction, linguistic competence, and even communities of computational agents. Epistemic logic, broadly construed, provides the tools to study these notions in a systematic manner. On the other hand, these various uses of epistemic logic have invited a number of extensions, e.g. from one to many agents, from statics to dynamics, which in turn have raised new questions for the very theory of knowledge and belief, where it all started.

We have been lucky enough to be able to gather interviews of both the pioneers who forged the discipline as well as a good sample of those working at its contemporary frontier.[1] Epistemic logic is still relatively young. Yet, the snapshot we get from this volume is one of a field that is reaching maturity without losing its dynamism nor its attractiveness. The new generation is standing on solid shoulders, which gives good hope for the future of epistemic logic.

Vincent F. Hendricks & Olivier Roy
Copenhagen and Groningen
August 2010

---

[1] The intersection of these two sets surely not empty!

Acknowledgements

We are particularly grateful to the contributors for devoting time to writing such erudite, enlightening and often thought-provoking interviews and grateful to the community in general for showing interest in this project. In addition we would like to express our gratitude to our publisher **Automatic Press ♦ ∨|P**, in particular senior publishing editor V.J. Menshy, for continuing to take on these 'rather unusual academic' projects.

This volume in the 5 *Questions Series* is dedicated to **ESSLLI** 2010—a good breeding ground for epistemic logic.

Vincent F. Hendricks & Olivier Roy
Copenhagen and Groningen
August 2010

1

Horacio Arló-Costa

Associate Professor
Carnegie Mellon University, USA

---

1. Why were you initially drawn to epistemic logic?

My initial interest in epistemic logic emerged with recent attempts to construct an epistemic semantics for modalities (and conditionals). This foundational program utilized an epistemologically viable alternative to possible worlds semantics. While most existing possible worlds semantics for unary and dyadic modals are often heavily influenced by grand metaphysical ideas (for example, those of Lewis and to some extent those of Kripke), the program of epistemic semantics tries to show that the usual metaphysical assumptions built into possible worlds semantics are dispensable. An example is the similarity relation used by Lewis in his semantics for counterfactuals. Many reacted unfavorably to the idea that the fundamental primitive that one has to assume to understand conditionals and important related notions like causal influence is that of distance between possible worlds. Is it possible to construct semantics for most of the unary and binary modalities without appealing to such abstractions, instead appealing to more terrestrial notions like entrenchment or epistemic value? The answer is Yes, although the epistemic program found its own internal puzzles, and in fact, there really is no single program of this kind but various foundational programs sharing a common epistemic flavor. For example, programs involving probabilistic semantics or belief revisionistic semantics abound. Initially I worked within the latter program, and recently I have worked in the former. Much of my work in the latter program focused on building epistemic models for binary modalities (conditionals). In [Arló-Costa, 1999] I offered a completeness result for auto-epistemic operators of the sort proposed by Isaac Levi in [Levi, 1997].

More recently I returned to epistemic logic, influenced by the foundational program for neighborhood semantics deriving from the work of Dana Scott and Richard Montague. Neighborhood semantics appeared initially as a formal alternative to Kripkean semantics, permitting a more flexible and encompassing study of modals. I published a paper [Arló-Costa, 2002] in which I proposed an epistemic interpretation of neighborhoods: the set of propositions in the neighborhood of a world w should be seen as the set of propositions known/presumed/supposed by an agent at w. I also extended this basic idea to the first order level. Later on Eric Pacuit and I [Arló-Costa & Pacuit, 2006] proved a series of general representation results for this type of semantics.

Neighborhood semantics can provide truth conditions for unary and binary models, and at the same time it can be seen as an epistemically motivated semantic program. My recent work shows that neighborhood semantics is the ideal tool for providing truth conditions for epistemic modals. The idea is to enrich the neighborhoods with structures like Grove systems or to define the neighborhoods probabilistically. This permits one to unify various tools that usually are applied separately: the tools of belief revision, probabilistic semantics, and neighborhood semantics. Paul Pedersen and I [Arló-Costa & Pedersen, 2010b] argued recently that this level of complexity is needed to understand the logical behavior of indicative conditionals (which are usually presented as typical examples of binary epistemic modals).

2. What example(s) from your work, or work of others, illustrates the relevance of epistemic logic?

Epistemic logic plays a crucial role in various branches of formal epistemology. For example, epistemic modalities are essential to disentangle various puzzles relating the notions of belief simpliciter and the notion of degrees of belief (to borrow from the title of a recent paper by Hannes Leitgeb [Leitgeb, 2009]). Henry Kyburg formulated a well-known puzzle— the so-called lottery paradox— relating the two notions. The moral of the paradox is that the notion of belief determined by an acceptance rule based on high probability should not be logically closed under conjunction. Initial diagnoses of the paradox lead in the wrong direction, suggesting that the use of an acceptance rule in terms of high probability is incompatible with the endorsement of classical logic (the introduction rule for conjunction fails). In [Arló-Costa, 2002], I argued that the paradox should be formulated in terms of a non-Kripkean epistemic modality for which the rule inferring Bel(p &

q) from Bel(p) and Bel(q) fails. This resolution of the paradox is compatible with the use of the classical notion of consequence and the classical Boolean operators. Kyburg and Teng endorsed this way of analyzing the paradox in a recent article [Kyburg & Teng, 2002]. In order to see the right solution it is essential to have the expressive power given by a classical modality (in the sense of 'classical' used by Chellas in his textbook [Chellas, 1980]). There is a rich connection between non-adjunctive logics and the corresponding classical modalities (see [Arló-Costa, 2005]).

Recent work connecting plain belief and degrees of belief ([Leitgeb, 2009], [Lin, 2009]) articulates the relation in a different way (logical closure need not be abandoned). Again it is essential to have unary and binary modalities in order to represent the right bridge rules connecting belief and degrees of belief (and conditional belief and conditional degrees of belief). Rohit Parikh and I [Arló-Costa & Parikh, 2005] recently offered a model reducing full belief to degrees of belief.

Deep relations between knowledge, truth, and self-reference are also revealed when one has at least the expressivity of an epistemic modality. The so-called paradox of the Knower, first formulated by Montague and Kaplan [Montague & Kaplan, 1960], presents an interesting conceptual puzzle and a formal challenge for the formulation of an adequate epistemic logic. Recently some variants of Fitting's Quantified Logic of Proofs (a first order extension of LP first formulated by Artemov) have been proposed in order to resolve these puzzles [Dean and Kurokawa, 2009]. The proposal is very interesting in many ways, but the paradox seems to persist even in this type of logic, which allows for a more fine-grained representation of knowledge than the epistemic logics deriving from the work of Hintikka [Arló-Costa & Kishida, 2009].

Various definitions of knowledge and truth-tracking depend on the use of binary modalities interacting with a doxastic operator. In general, the use of logical environments where one can allow for the simultaneous representation of knowledge, belief, and time are critical not only for purely epistemological reasons but also for applications in many fields outside of philosophy (see next section for examples).

3. What is the proper role of epistemic logic in relation to other disciplines, for instance mainstream epistemology, game theory, computer sciences or linguistics?

I mentioned some relations with mainstream epistemology in the previous section. Some deep problems in the theory of knowledge and belief, concerning the relations of knowledge with truth and self-reference and belief with degrees of belief are only evident when one has the proper logical tools at one's disposal. Aside from this, there are many foundational problems related to the interpretation and philosophical understanding of epistemic logics. These problems do not only pertain to philosophical logic but to a discipline that today is usually called formal epistemology. I see formal epistemology not only as a collection of formalisms but also as a discipline capable of reflecting philosophically about the applicability and adequacy of these formalisms for the solution of a variety of problems inside and outside philosophy.

Epistemic logic and belief revision have been recently applied in game theory, or more broadly, in the discipline that Robert Aumann calls "interactive epistemology." This epistemic program in the foundations of game theory has used a variety of logical tools, and among them epistemic logics have played a prominent role. Robert Aumann developed his own version of epistemic operators endowed with its own semantics and used them to prove interesting and deep theorems, like the one that shows that it is not possible to agree to disagree [Aumann, 1976]. More generally Aumann has used what is essentially epistemic logic in order to provide epistemic foundations for some of the central notions of equilibrium (Nash equilibrium, subgame perfect equilibrium, and correlated equilibrium). So the importance of epistemic logic in this area is clear. But it is probably true that the epistemic program is game theory is only in its infancy. For example, it seems that the epistemic program could be used in order to develop models of bounded agents playing in real conditions. Now we have a relatively well-developed branch of behavioral game theory that gives basic data about real play. For example, McKelvey and Palfrey [McKelvey & Pafrey, 1992] report about a game where two players alternately get a chance to take the larger portion of a continually escalating pile of money. As soon as one person takes, the game ends with that player getting the larger portion of the pile, the other player getting the smaller portion. If one views the experiment as a complete information game, all standard game theoretic equilibrium concepts predict the first mover should take the large pile on the first round. The experimental results show that this does not occur. We have today sophisticated epistemic models of centipedes of this type. Is it possible to use them to produce cognitive models of real play? There are interesting cognitive models of this type already [Camarer, Ho and Chong, 2004] but they make heavy use of probability hierarchies. In a game like the centipede a purely qualitative model should suffice. To develop these types of models one needs epistemic logics that, in turn, are capable of representing real players (and that therefore are not affected by the problem of logical omniscience – see below for comments about logical omniscience). Sergei Artemov and collaborators at CUNY and Alexandru Baltag and collaborators in Oxford have done recent work in this direction. Much of the work of Baltag appeals to Dynamic Epistemic Logic (DEL) a variety of epistemic logic mainly investigated in Amsterdam by Johan van Benthem and associates.

Epistemic logic has been applied to many branches of computer science. The work of Fagin, Halpern, Moses and Vardi [Fagin et al., 1995] has many applications beyond the boundaries of computer science. For example they offer a complete axiomatization of the notion of common logic in a standard epistemic logic, showing that it is indeed possible to reason about common knowledge in a feasible manner. This work is a typical example of the advantage of using the tools of epistemic logic to reason about common knowledge (Lewis initial definition did not appeal to epistemic operators and remains unclear in many ways).

The program of model checking developed at CMU has made abundant use of epistemic and temporal logics. The problem in model checking runs as follows: given a model of a system, test automatically whether this model meets a given specification. Both the model of the system and the specification are formulated in a logical formalism. The main logical problem consists in checking whether a given structure (which could be a Kripke structure for an epistemic logic) satisfies a given formula. Clarke, Emerson, and Sifakis shared the 2007 Turing Award for their work on model checking (see [Clarke et al., 1999]).

Finally, there is an interesting connection between game theory and pragmatics and semantics. For example, there are intriguing game theoretical models of Grice's notion of implicature. This is a way of seeing interactive epistemology as a broad discipline with applications in linguistics, computer science and economics.

In philosophy there is a fair amount of recent interest in the study of epistemic modals. I mentioned some of this work above. An interesting reference is [Kolodny and MacFarlane, 2008]. This paper offers truth conditions for indicatives (a typical case of a binary epistemic modal). A criticism of this account appears in [Arló-Costa & Pedersen, 2010b].

4. Which topics and/or contributions should have had more attention in late 20th century epistemic logic?

Although Hintikka's book (Knowledge and Belief) did not limit itself to the study of propositional epistemic logics (various references to first order extensions are made), most of the mainstream work in epistemic logic in the 20th century limited itself to the consideration of propositional formalisms. It is obvious that this is a severe limitation. Perhaps the influence of computer science was decisive regarding this bias. First order extensions of binary epistemic modals are practically unknown in spite of their intrinsic interest. Perhaps the paper [Arló-Costa & Pacuit, 2006] is the first systemic attempt to study a large family of first order classical modalities in the literature. But this work is relatively recent and it belongs to the current century.

Perhaps the most important neglected problem in epistemic logic even today is that of logical omniscience. This is a problem with both logical and philosophical ramifications that affects practically all our common representations of knowledge and belief in Bayesian epistemology. There is a recent wave of work in this area, but perhaps we do not yet have a completely convincing solution or family of solutions. The construction of credible cognitive models of real play in game theory depends on having logics that are not affected by the problem of logical omniscience. The so-called classical epistemic logics below K are not affected by the most common forms of logical omniscience (closure under known implication, closure under conjunction, for example). But the formalism presupposes that if p is logically equivalent to q then Bel(p) is equivalent to Bel(q), which constitutes in itself a form of logical omniscience.

Still, perhaps what is needed is a philosophical diagnosis of the problem, which, in turn, requires an evaluation of the interface between the normative and the descriptive.

Logic is usually associated with normative ideals. Is it possible to have logics that are flexible enough to describe the behavior of real agents? This depends on having a clear view about the program of bounded rationality as first formulated by Simon. And this remains to a large extent an open problem both formally and conceptually.

5. What are the most important open problems in epistemic logic and what are the prospects for progress?

One interesting open problem (about which we are making good progress) is to investigate the connections between belief, conditional belief and degrees of belief. This is a crucial problem for formal epistemology that connects two central doxastic models, one qualitative and another probabilistic. One obvious extension of this program is to find the right connections between belief change and the usual models of probability kinematics. These are well-formulated problems that will probably be fully solved during this decade.

The aforementioned type of research links the Hintikka-type approach to epistemic logic with work in Bayesian epistemology. It would be interesting to explore as well the connections of formal learning theory and epistemic logic. My colleague Kevin Kelly and Vincent Hendricks in Denmark have done preliminary work in this direction.

Another tractable problem consists in the developing of first order extensions of our models of belief and conditional belief and belief change. Again this is an area where progress is guaranteed to occur soon.

The problem of logical omniscience, its solution or dissolution, remains as a stubborn problem affecting most of epistemic logic and Bayesian representations of belief. This problem is connected intimately with the problem mentioned above related to the formulation of realistic theories of actual knowledge. As I explained above, perhaps the problem of logical omniscience is a family of problems rather than a single problem. At least it appears under different disguises in various areas of formal epistemology. This problem or family of problems is without a doubt harder than the aforementioned puzzles. Perhaps the first steps towards its solution consist in developing credible cognitive models of real play in game theory.

Another interesting and tractable area is the study of 'fast and frugal' heuristics. This was the strategy used by Simon in order articulate a project of bounded rationality. This is an area that begs for clear mathematical modeling. One can see a heuristic as an input-output mechanism implementing a selection function (i.e., also called a "choice" function). What are the properties of such a selection function for different formulations of the heuristic? Are there heuristics that implement 'rational' selection functions (with properties like alpha, etc)? If so, this is tantamount to having bounded methods capable of approximating ideals of rationality. And when some crucial properties of rationality fail (for example, when there are cycles and failures of transitivity in the underlying preference relation), is it still possible to characterize fully the corresponding heuristics? Preliminary work in this direction is presented in [Arló-Costa & Pedersen, 2010a], but much work remains to be done in this area.

References

[Arló-Costa, 1999] Qualitative and Probabilistic Models of Full Belief, Proceedings of *Logic Colloquim'98, Lecture Notes on Logic* 13, S. Buss, P.Hajek, P. Pudlak (eds.), ASL, A. K. Peters, 1999.

[Arló-Costa, 2002] H. Arló-Costa, First order extensions of classical systems of modal logic: The role of the Barcan schemas, *Studia Logica* 71: 2002, 87-118.

[Arló-Costa, 2005] H. Arló-Costa, Non-Adjunctive Inference and Classical Modalities, *Journal of Philosophical Logic*, vol. 34, no5-6, 2005, 581-605.

[Arló-Costa & Parikh, 2005] H. Arló-Costa and R. Parikh, Conditional Probability and Defeasible Inference, *Journal of Philosophical Logic*, 34, 2005, 97-119.

[Arló-Costa & Pacuit, 2006] H. Arló-Costa and E. Pacuit, First-Order Classical Modal Logic, *Studia Logica*, 84, 2, 2006, 171-210.

[Arló-Costa & Kishida, 2009] H. Arló-Costa and K. Kishida, Three proofs of the Knower in the Quantified Logic of Proofs, 2009, *FEW'09*.

[Arló-Costa & Pedersen, 2010a] H. Arló-Costa and A. P. Pedersen. Fast and Frugal Heuristics: Take The Best and the Priority Heuristics in Perspective, 2010. *Synthese Conference on Epistemology and Economics.*

[Arló-Costa & Pedersen, 2010b] H. Arló-Costa and A. P. Pedersen, Indicative Conditionals, Obligation and Decision Theoretic Conditionals, 2010, *FEW'10*.

[Auman, 1976] R. Aumann, Agreeing to Disagree, *Annals of Statistics*, 4, 1976, 1236-1239.

[Camarer et al., 2004] C.F. Camarer, T-H Ho & J.K. Chong, A cognitive hierarchy model of games, *The Quarterly Journal of Economics*, 2004, 861-898.

[Chellas, 1980] B. F. Chellas. *Modal logic: an introduction.* Cambridge University Press, Cambridge, 1980.

[Clarke et al., 1999] E. M. Clarke, Jr., O. Grumberg and D. A. Peled, *Model Checking*, MIT Press, 1999.

[Dean & Kurokawa, 2009] W. Dean and H. Kurokawa, *Knowledge, Proof and the Knower*, 2009, FEW'09.

[Fagin et al., 1995] R. Fagin, J.Y. Halpern, Y. Moses and M.Y. Vardi, *Reasoning About Knowledge*, MIT Press, 1995.

[Kahaneman, 2002] D. Kahneman, Maps of Bounded Rationality, Nobel lecture, 2002.

http://nobelprize.org/mediaplayer/index.php?id=531

[Kolodny and MacFarlane, 2008] N. Kolodny and J. MacFarlane, *Ifs and Oughts*, manuscript, University of California, Berkeley, Draft of August 11, 2008, submitted to the Journal of Philosophy.

[Kyburg & Teng, 2002] H. E. Jr. Kyburg and C. M. Teng, *The Logic of Risky Knowledge*, proceedings of WoLLIC, Brazil, 2002.

[Leitgeb, 2009] H. Leitgeb, Reducing belief simpliciter to degrees of belief, manuscript, Bristol University, 2010.

[Levi, 1997] I. Levi, The Logic of Full Belief, in *The Covenant of Reason: Rationality and the Commitments of Thought*, Cambridge University Press, Cambridge, 1997, 40-70.

[Lin, 2009] H. Lin, A logical theory of belief without certainty, manuscript, CMU, 2010.

[McKelvey & Pafrey, 1992] R. D. McKelvey & T.R. Palfrey, An experimental study of the centipede game, *Econometrica*, 60, 4, 1992, 803-836.

[Montague & Kaplan, 1960] R. Montague and D. Kaplan. A paradox regained, *Notre Dame Journal of Formal Logic*,1, 1960, 79-90.

2

Sergei Artemov

Distinguished Professor

The Graduate Center of the City University of New York, USA

---

1. Why were you initially drawn to epistemic logic?

I have always been attracted to foundational issues. My teenage interest in the sciences and technology resulted in Moscow University training in mathematics. From the mathematical subjects offered, I chose logic and foundations. My first conference talk and publication were in set theory. However, inspired by my mentor A.N. Kolmogorov's example, I shifted my focus to constructive logics and the theory of proofs. Provability has fascinated me both as a general intellectual phenomenon and in connection with constructive/intuitionistic logic via Brouwer-Heyting-Kolmogorov semantics. This determined my interest in modal provability logic in which I have worked actively since 1979 (cf. [5, 10, 12]).

In 1991 I came to the conclusion that provability logic models only a fragment of the provability phenomenon, that mostly related to the implicit representation of proofs by the existential quantifier over proof codes in first-order logic. Due to the E-sickness of first-order logic (cf. [1, 5]), this model deviates considerably from the intended intuitive properties of provability, e.g., reflexivity 'provable yields true.' In particular, the principal provability logic system **GL** is not compliant with **S4**, which Gödel endorsed in 1933 and 1938 as the modal logic of provability [14, 15].

Subsequently, in 1994 I developed an alternative logical system for provability which is now known as the Logic of Proofs **LP**. The Logic of Proofs makes proof terms the principal objects and provides an alternative model of provability which meets Gödel's specifications.

From the very beginning, the Logic of Proofs has been seen as a general mathematical model of evidence, e.g., proofs. Since the end of the 1990s, the Logic of Proofs project has evolved into a broader subject, justification logic, in which many researchers are now making principal contributions: Mel Fitting, Vladimir Krupski, Roman Kuznets, Robert Milnikel, Elena Nogina, Brian Renne, Tatiana Yavorsakaya, and many others.[1] I would like to think that justification logic is becoming an integral part of epistemic logic.

I now see great potential for epistemic logic as the foundation of other disciplines such as epistemic game theory, as well as to the area of information control and security.

2. What example(s) from your work, or work of others, illustrates the relevance of epistemic logic?

I will provide two examples from my most recent work on epistemic game theory.

Though game theory acknowledges the importance of epistemic states of players, the standard textbooks in game theory, e.g., [18], when considering motivational examples of 2×2 strategic games (such as the Prisoner's Dilemma (PD), Bach or Stravinsky (BoS), Hawks and Doves, Matching Pennies, etc.) avoid considering games in which epistemic states of players matter. In particular, in PD, each player has a dominant strategy to confess, and players' knowledge about opponents' intentions are irrelevant. In BoS, no player has a dominant strategy and hence all their choices are Aumann-rational: under these conditions, an additional epistemic assumption of knowledge of rationality does not change the predicted behavior of players either. All five introductory examples in [18], pp. 15–17, are of such a nature: they are 'epistemically neutral,' i.e., additional knowledge of other players' rationality does not change the way the game is played. However, not all strategic 2×2 games are epistemically neutral. Consider the following War and Peace dilemma (WP) first introduced in [3].

> Imagine two neighboring countries: a big, powerful B, and a small S. Each player has the choice to wage war or keep the peace. The best outcome for both countries is peace. However, if both countries wage war, B wins easily and S loses everything, which is the second-best outcome for B and the worst for S. In situation $(war_B, peace_S)$, B loses internationally, which

---

[1] The milestone paper [13] provided a necessary semantical treatment of the Logic of Proofs by introducing Kripke-Fitting models, which became the standard tool in this area.

is the second-best outcome for S. In $(peace_B, war_S)$, B's government loses national support, which is the worst outcome for B and the second-worst for S.

	war_S	$peace_S$
war_B	2,0	1,2
$peace_B$	0,1	3,3

There is one Nash equilibrium, $(peace_B, peace_S)$, consisting of the best outcomes for both players. It might look as though rational players should both choose *peace*. However, such a prediction is not well-founded unless certain epistemic conditions are met.

If B is not aware of the other player's rationality, S chooses *peace* but B considers either choice by S possible and, according to Harasnyi's rationality postulates (cf. [16]), chooses the maximin solution, i.e., *war*. If S's rationality is known to B, then both players choose *peace*. So in the War and Peace dilemma, despite the fact that

a) for both countries, the best choice is *peace*;
b) it is the only Nash equilibrium in the game;
c) both countries behave rationally;

to secure the Nash equilibrium outcome, an additional epistemic condition should be met, e.g., the big country should know that its small neighbour will behave rationally.

The reason why WP is epistemically sensitive is that one of the players (small country) has a dominant strategy whereas the other player (big country) does not. It is easy to see that 50% of all generic strategic 2×2 games with ordinal payoffs are of this sort, i.e., are epistemically sensitive. This observation illustrates the potential of epistemic logic in game theory: any responsible analysis of epistemically sensitive games (and about half of the most common strategic 2×2 games are of this kind) should be based on epistemic logic.

The second example compared probabilistic methods of dealing with epistemic matters in game theory with those provided by epistemic logic. A classical paper [9] of epistemic game theory established that

> in a strategic game with Aumann-rational players, for any Nash equilibrium σ, there is a belief system in which each player assigns probability one to σ.

When analyzing the same setup by logical methods, [4] states that

> *in a strategic game with Aumann-rational players, for any Nash equilibrium σ, it is consistent with the rules of the game to assume that all players know that σ is played*

which immediately yields the following impossibility corollary:

> *If a strategic game with Aumann-rational players has two or more Nash equilibria, then such a game does not have a definitive solution that players can derive from the game description.*

Note that Nash in [17] based his equilibrium solution concept on the assumption that the game has a definitive solution that players can derive from the game description.[2] The aforementioned corollary shows that games with more than one Nash equilibrium do not have definitive solutions in pure strategies within the scope of Aumann rationality.

3. What is the proper role of epistemic logic in relation to other disciplines, for instance mainstream epistemology, game theory, computer sciences or linguistics?

Epistemic logic has played a prominent role in computer science and artificial intelligence. This is a well-covered classical topic.

Epistemology as a philosophical discipline is heavily influenced by logical language and methods. There now exists a quite potent logical framework which naturally accommodates such fundamental epistemic notions as knowledge, belief, and justification. In particular, in [2], the paradigmatic Gettier example has been formalized in quantifier-free justification logic. The corresponding analysis has revealed hidden assumptions and redundancies. Mainstream epistemologists often use formal or semiformal logical reasoning in their studies, which could benefit greatly from a more systematic connection to epistemic logic. However, epistemic logic has yet to earn the trust it deserves there, to convince mainstream

---

[2] Many thanks to Adam Brandenburger for attracting attention to this matter.

epistemologists that formal logical methods can be a useful tool, and that it is worth studying and applying.

Game theory has long studied epistemic conditions under which classical game-theoretical solution concepts, such as backward induction, work in the intended way (cf. Aumann's Theorem on Rationality [8]). Game theory is now embracing a broader view of games, which includes specifying epistemic states of players. This opens a whole new frontier in game theory in which epistemic logic is destined to play an essential role. The current state of game theory provides an historical opportunity for epistemic logic to rise above the level at which logic formalizes existing concepts and studies sufficient epistemic conditions. It is time for epistemic logic to lead and offer principally new concepts for mainstream game theory, especially in its new domain, epistemic game theory, which considers arbitrary epistemic states of players rather than the standard (and often unrealistic) assumption of common knowledge of rationality. Knowledge-based rational (KBR) solutions in games of perfect information [3] can serve as an example. KBR naturally absorbs classical solution concepts such as Nash equilibria and subgame perfect equilibria, backward induction, and maximin and provides a definitive (unique) solution for each generic game of perfect information: no other existing method achieves this. Moreover, KBR is based on classical game-theoretical principles of rationality as presented by Aumann ([8]) and Harsanyi ([16]): epistemic logic effectively helps to make proper use of these and other classical game-theoretical assumptions.

4. Which topics and/or contributions should have had more attention in late 20th century epistemic logic?

Constructive methods have been underdeveloped and underutilized in 20th century epistemic logic. Mathematical logic has achieved a certain (though not always properly acknowledged) balance between proof-theoretical and model-theoretical methods.

Gödel's Completeness Theorem states that valid formulas of first-order logic are exactly those derivable in the corresponding proof system with a decidable set of axioms and rules. Gödel's Incompleteness Theorems can be understood via nonstandard models of arithmetic. Compactness considerations which are instant in the proof-theoretical framework are powerful tools in many model-theoretical tasks, etc.

However, 20th century epistemic logic has been strongly dominated by model methods, specifically Kripke models and their numerous variations. Perhaps proof-theoretical methods are more technically involved and require more work to muster. This usually does not create a problem for a mathematically trained logician whereas epistemic logicians with other backgrounds, e.g., philosophical, engineering, or computer science, might feel tempted to cut corners and resrict their technical toolkits to less sophisticated, purely model-theoretical (mostly Kripke-style) analysis of epistemic problems. Such a one-sided approach very often works just fine. However, in some remarkable situations, it misses the mark. Let us consider some examples.

The logical omniscience problem in epistemic logic emerges due to the unbounded epistemic closure principle

$$\Box(A \to B) \land \Box A \to \Box B \tag{2.1}$$

according to which an agent's knowledge is closed under logical deduction. By this principle, the agent knows all deductive consequences of his/her assumptions which is, of course, unrealistic, e.g., when the number of steps in the deduction becomes large. This aspect of the logical omniscience problem gets fair treatment by taking into account the internalized deduction complexity (cf. [6, 7]). Epistemic assertion $\Box F$ is treated constructively as supported by a corresponding justification term t (a footprint of a formal derivation, or a proof); $t:F$ stands for 't is a justification for F.' The epistemic closure principle in such explicit form becomes

$$s{:}(A \to B) \land t{:}A \to (s{\cdot}t){:}B \tag{2.2}$$

where st is an 'application' of s to t, i.e., a justification term which combines s with t and performs the 'atomic' action of deducing B from $A \to B$ and B. Naturally, the complexity of $s{\cdot}t$ exceeds the sum of complexities of s and t. Even such a crude mechanism of evidence tracking suffices to rule out epistemic assertions that require 'too large' justifications: it is not feasible to even formalize such assertions in the explicit epistemic language. Such a complexity treatment of logical omniscience becomes possible only when we consider the proof-theoretical side of epistemic logic that was missing in the traditional Kripke-style analysis.

Another example concerns such foundational issues as the epistemic closure principle (2.1) itself. Kripke's well-known Red Barn Example (RBE) gives an epistemic scenario in which (2.1) is violated (cf. [2]). In particular, RBE provides a plausible description which involves three belief assertions appearing as the results of unrelated observations $\Box(A \to B)$, $\Box A$ and $\Box B$ such that the first two amount to knowledge whereas the third does not. The reason for this failure of epistemic closure is that in Boolean logic with material implication '$\to$,' principle (2.1) in fact rules out only truth-value combinations under which $\Box(A \to B)$ and $\Box A$ hold but $\Box B$ does not. In particular, according to (2.1), whenever beliefs $\Box(A \to B)$ and $\Box A$ are knowledge, then belief $\Box B$ is also knowledge regardless of its origin, even if it is unrelated to $\Box(A \to B)$ and $\Box A$. *RBE* provides a clever example of why (2.1) as it stands can fail.

The explicit version (2.2) saves the epistemic closure principle, at least from this kind of attack. It postulates knowledge of B, given knowledge of A and $A \to B$, only when the justification of knowledge assertion of B is related to justifications for A and $A \to B$ in a certain regular way.

This immediately resolves the Red Barn puzzle and provides insight as to why the pure modal version (2.1) of the epistemic closure principle has widely been assumed to be true. People tend to read material implication in a constructive way by assuming that $\Box B$ in (2.1) is somehow related to $\Box(A \to B)$ and $\Box A$. This is how (2.2) treats epistemic closure, but this is not what the purely modal version (2.1) of epistemic closure actually states. By the realization theorems of justification logic (cf. [2]), any valid statement of epistemic logic in fact has an exact explicit reading which corrects the aforementioned defects of material implication. It is perhaps this explicit reading which we intuitively assign to modal epistemic principles that makes the latter seem plausible.

Note that now, as justification logic fills the proof-theoretical void in epistemic logic, new exciting areas such as the knowledge vs. justified true belief discussion, Russell and Gettier-style examples, etc., are within reach of formal logical methods. Formal methods do not usually resolve fundamental philosophical issues but they provide the necessary reliable framework for more advanced reasoning about these issues.

5. What are the most important open problems in epistemic logic and what are the prospects for progress?

These problems are of a mostly methodological character. Epistemic logic has to become more balanced, e.g., absorb constructive proof-theoretical methods externally (proof systems) and internally (proof terms), and adjust the model theory respectively: Kripke-Fitting models (cf. [2, 13]) provide a sense of direction here. Epistemic logic has to develop and study evidence-tracking systems; there is a definite interest in this kind of machinery from many areas of potential application. Epistemic logic has to incorporate 'active agency' and sort out belief revision. Most importantly, epistemic logic has to actively expand to other areas: game theory, information security, ontological components of data bases, mainstream epistemology, and law are immediate examples.

In particular, epistemic game theory (cf. [11], and Brandenburger's paper in this volume) traditionally models players' beliefs by assigning probabilities to epistemic states. However, the methods of epistemic modal logics can also be productive and provide new insights there as well. These considerations prompt a foundational problem of reconciling probabilistic and logical approaches to beliefs.

2.1 REFERENCES

[1] S. Artemov. Explicit provability and constructive semantics. *Bulletin of Symbolic Logic*, 7(1):1–36, 2001.

[2] S. Artemov. The Logic of Justification. *The Review of Symbolic Logic*, 1(4):477-513, 2008.

[3] S. Artemov. *Knowledge-based rational decisions*. Technical Report TR-2009011, CUNY Ph.D. Program in Computer Science, 2009.

[4] S. Artemov. *The impossibility of definitive solutions in some games*. Technical Report TR-2010003, CUNY Ph.D. Program in Computer Science, 2010.

[5] S. Artemov and L. Beklemishev. Provability logic. In D. Gabbay and F. Guenthner, editors, *Handbook of Philosophical Logic, 2nd ed.*, volume 13, pages 189–360. Springer, Dordrecht, 2005.

[6] S. Artemov and R. Kuznets. Logical omniscience via proof complexity. In *Computer Science Logic 2006*, volume 4207, pages 135–149. Springer Lecture Notes in Computer Science, 2006.

[7] S. Artemov and R. Kuznets. Logical omniscience as a computational complexity problem. In *Aviad Heifetz, editor, Theoretical Aspects of Rationality and Knowledge, Proceedings of*

the Twelfth Conference (TARK 2009), Stanford University, California, July 6-8, 2009. ACM. , pages 14–23, 2009.

[8] R. Aumann. Backward Induction and Common Knowledge of Rationality. *Games and Economic Behavior*, 8:6–19, 1995.

[9] R. Aumann and A. Brandenburger. Epistemic conditions for Nash equilibrium. *Econometrica: Journal of the Econometric Society*, 64(5):116–1180, 1995.

[10] G. Boolos. *The Logic of Provability*. Cambridge University Press, Cambridge, 1993.

[11] A. Brandenburger. The power of paradox: some recent developments in interactive epistemology. *International Journal of Game Theory*, 35:465-492, 2007.

[12] D. de Jongh and G. Japaridze, *Logic of provability*, In: Handbook of Proof Theory, S. Buss (Ed.), Elsevier, pp. 475–546, 1998.

[13] M. Fitting. The logic of proofs, semantically. *Annals of Pure and Applied Logic*, 132(1):1–25, 2005.

[14] K. Gödel. Eine Interpretation des intuitionistischen Aussagenkalkuls. *Ergebnisse Math. Kolloq.*, 4:39–40, 1933. English translation in: S. Feferman et al., editors, *Kurt Gödel Collected Works, Vol. 1*, pages 301–303. Oxford University Press, Oxford, Clarendon Press, New York, 1986.

[15] K. Gödel. Vortrag bei Zilsel/Lecture at Zilsel's (*1938a). In Solomon Feferman, John W. Dawson, Jr., Warren Goldfarb, Charles Parsons, and Robert M. Solovay, editors, *Unpublished essays and lectures*, volume III of *Kurt Gödel Collected Works*, pages 86–113. Oxford University Press, 1995.

[16] J.C. Harsanyi. Games with Incomplete Information Played by "Bayesian" Players, I-III. *Management Science*, 14:159-182, 320-334, and 486-502, 1967/68.

[17] J. Nash. *Non-Cooperative Games*. Ph.D. Thesis, Princeton University, Princeton, 1950.

[18] M. Osborne and A. Rubinstein. *A Course in Game Theory*. The MIT Press, 1994.

3
Robert Aumann

Professor of Mathematics
Center for the Study of Rationality
and the Department of Mathematics
The Hebrew University of Jerusalem, Israel

Interviewed on January 23rd, 2010, in Bohn, Germany. The editors would like to thank Prof. Dr. Hildenbrand for his hospitality on this occasion.

O.R.:
Professor Aumann, many thanks for accepting our invitation to give us an interview. First question: What initially drew you to epistemic logic?

R.A.:
It was a natural outgrowth of work in game theory. The beginning came with work on common knowledge, the Agreement Theorem. This states that when two people have common priors and common knowledge about each other's probability assessments, then these probability assessments must be the same. The moment you start talking about common knowledge you get drawn into all kinds of epistemic considerations. The question is, what led to that? What led to the interest in common knowledge?

I wrote a paper in the very early '70s called 'Subjectivity and Correlation in Randomized Strategies'.[1] Randomised strategies go back to von Neumann and Morgenstern. These are mixed strategies based on some kind of coin toss, roll of the dice or spin of the roulette wheel. On the other hand there was the work of Savage, going back to Ramsey and de Finetti. This concerned *subjective*

---
[1] *Journal of Mathematical Economics*, 1(1), pp. 67-96, 1974.

probabilities, which started to bloom in the '50s and early '60s. So the next step seemed obvious: what happens when you try to apply subjective – rather than objective – probabilities to randomisation? This question led to the paper about subjectivity and correlation in randomised strategies, which was also the beginning of *correlated* equilibria. In this paper, I show that you can have subjective mixing and that subjective mixing leads to different equilibria, called in the paper *subjective* equilibria. You even get different pure equilibria from those you get with just objective mixing. At the end of the paper I discuss how this can occur. Why would people have different subjective evaluations? At that time it was an open question.

One summer I was at the Stanford Institute for Mathematical Studies in the Social Sciences. Sitting with Kenneth Arrow and Frank Hahn in Frank's office, we were discussing the question and I recall that the idea somehow arose from that discussion. I went back to my office and started thinking about it and that's how the agreement theorem was born. At the time it actually seemed too simple to publish. As you know, the proof only takes two lines; but I think the main thing is the *formulation* of the idea: it's not the proof itself. The proof seemed so simple that it somehow seemed beneath my dignity to publish it. I explained the theorem to Arrow and Hahn, and at first they didn't want to believe it. Finally they became convinced, and then actually convinced me that this was worth publishing. It became one of the two papers in my oeuvre that are cited most widely, and that was the beginning of my interest in epistemic logic.

Then one tries to develop, to get at the foundations, and this is how I became especially interested in the *syntactic* part. One question that is very puzzling to people who read the Agreeing to Disagree paper is: where does the partition *come from*? Somehow the knowledge partitions are assumed to be known to the players. In fact they're *common* knowledge to the players. This is built into the system. Where does it come from? These are the questions that most often arise with regard to that paper. How do people know each other's partitions? Why is this sort of given? Why is the partition a *given*? And that got me interested in the syntactic angle. In some sense, when you talk about partition spaces, when you talk about a semantic representation in epistemic logic, you want to say: Hey, let's put it in English! You are making some kind of assertion about agreement or whatever it is that you're talking about. Establish this assertion and don't give me partition spaces and state spaces and stuff like that. Say it in English! And that's what the syntactic approach is all about: saying it in English.

I became obsessed with the idea of *deriving* a semantic representation from *plain English*, which is what syntactics is all about. 'Syntactic epistemic logic' is just saying it in English. Dov Samet, an outstanding expert in epistemic logic, especially interactive epistemic logic, who did a PhD with me, put it this way, very picturesquely: If you want to explain it to your mother, you had better say it syntactically. Your mother won't understand if you try to say it semantically.

That sounds right. In semantics there's more grease, which allows you to work more easily. But you don't know where the foundations come from. Syntactic logic is more awkward, but it is simply logic. You derive one thing from another. You have certain axioms and you can argue about the axioms if you wish, but at least one thing follows from another. You don't have these sets and states of the world. What is a state of the world, anyway?

This is where the idea arose to view the semantic approach as *deriving* from the syntactic approach. You build a semantic approach *from* the syntactic approach.

This is how the whole thing started. I never see something and say: hey, this looks interesting, let's work on that. That's not the way it works. Rather, you have some problem and you say: we need to develop a tool to solve it. It is not 'this looks interesting and let's work on it'. People often ask me, why did I go into game theory? Well, I was faced with a problem, a specific problem that called for the use of game theory. So I said 'We have to study games.' It was a very specific problem. It's not like when you're a student at the university, you say: this subject looks attractive, that subject looks attractive, let's go into that subject. At least it was never that way in *my* research. Research is always very specifically purpose-oriented; one thing leads to another.

O.R.:
How did the semantic representation of epistemic logic come about, at the very beginning, at the origin of the Agreeing to Disagree paper?

R.A.:
I don't remember *exactly* the genesis of it, but it comes from Harsanyi's work. A type structure is essentially a partition structure. I think Harsanyi didn't think of it in *quite* that way, in terms of a state space with a partition. I don't know exactly how I came across that. I think it's already in the paper about subjec-

tivity and correlation, in the first issue of JME.[2] So I think it was '72, or thereabouts, fooling around with the Harsanyi type-space structure, that led to the structure of states and partitions. That's where it comes from; it's already in there. 'Agreeing to Disagree', as I said, sort of grew out of the problem left open in that paper about subjectivity and correlation. So it was a natural progression. It's a *restatement* of the Harsanyi structure. Not a very deep restatement but some sort of restatement. A type determines what each person thinks about the other person's types and payoffs. So everything is an n-tuple of types, and a slight restatement gives you a partition structure.

O.R.:

After Agreeing to Disagree, there's been a convergence between the work that you initiated on the semantic side about partition structures, and the work that was done by David Lewis in philosophy, for example. How did this convergence come about?

R.A.:

The David Lewis matter is really very, very interesting. I'll tell you my side of the story.

I wrote this paper and called this concept 'common knowledge'. This was published in 1976. Now around 1979 I ran across a paper in a philosophical journal which quoted my paper, I think. I'm not sure, though. It certainly quoted Lewis's book *Convention*,[3] which was published in '69. Perhaps it discussed a citation from my paper, too: I don't remember now, but it certainly quoted Lewis's book. I opened my eyes and said: 'Hey! What's going on here? This chap had the concept of common knowledge already in '69?' And, yes, it turned out that he did! I went back and bought the book or took it out of the library, and there it was. In 1969! And the amazing thing was that we used the *same word* for it, the same word that I thought I had invented. So from then on, when I wrote about this I started quoting Lewis.

Now, a year or two later Lewis sent a letter to the provost of Princeton University. In his PhD thesis, Sergio Ribeira DaCosta-Werlang had quoted me on common knowledge. Somehow the thesis was passed on to David Lewis as a reader. Lewis was furious because the thesis didn't quote him at all. It made believe that common knowledge was my concept. So I think Lewis wrote a let-

---

[2] *Journal of Mathematical Economics*, 1(1), pp. 67-96, 1974.
[3] *Convention: A Philosophical Study*, Cambridge, MA: Harvard UP, 1969.

ter to the provost in which he said ... well, he complained about this in fairly strong language. I don't remember the exact language. He sent a copy to Hugo Sonnenschein and a copy to me. So I got this copy and immediately wrote back saying: 'you know, Prof. Lewis, this is your concept. No question about it. I was not *aware* of your contribution to this when writing the '76 paper. I simply did not know. You might find this difficult to believe because we use the *same word*, we use the word 'common knowledge'. But it's true. And as soon as I became aware, which is a year or two ago now, I started citing you. Please accept my abject apologies. I'm not even saying that this is independent work. You can talk about independent work when you're talking about something that is done at approximately the same time or even a year or two later, but not when there's a hiatus of seven years. Because, you know, people talk at lunch or they talk in seminars and things filter through, very often without attribution. Both the idea itself and the name could have filtered through somehow without my being aware of it. So I'm not claiming independence. It's your contribution. The idea of common knowledge is your contribution and there's no question about it. This is all yours. I've been saying it for a while now, and I'll say it again. And let's meet.'

I was in Stony Brook at that time. I went down to Princeton, and I met Lewis and Kripke. We had a nice conversation. By that time there was a kosher dining club in Princeton. I'll have to tell you some stories about Princeton when I was originally there in the fifties. It was a hotbed of anti-Semitism. But you know, the world progresses and by 1981–82 there was a kosher dining club and in fact Hugo Sonnenschein became provost of Princeton. He's Jewish, of course. And then afterwards there was even a Shapiro who became President of Princeton University. He is also Jewish. So we're making progress. I had a nice day in Princeton and made it up with Lewis and Kripke, and everything was fine. And at every opportunity, including this one, I've been acknowledging that the idea of common knowledge is Lewis's idea. Period. No independence, nothing.

However, let me just add this: Lewis had the idea of common knowledge, and this is very clear. What he did *not* have is the agreement theorem. He did not have the agreement theorem, he had no glimpse of the agreement theorem, nothing like that. So I still lay claim to the agreement theorem.

O.R.:
You met both Lewis and Kripke. Were you already interested at

that time in the syntactic part of epistemic logic?

R.A.:
I'm not sure I got into the syntactic thing at that point. No, I don't think so, no. I was bothered by this question of where the partition structure comes from. That was in the early '80s: it bothered me even then. But I sort of finessed it. I said, well, *descriptively*, not *formally*: a state of the world includes everything that is relevant about that state, including what people know about that state. It's rather like a dictionary. It's nothing substantive, it's like a dictionary. And that statement occurs in my '87 paper 'Correlated Equilibria as an Expression of Bayesian Rationality'.[4] But actually to *describe* that dictionary, to do that and carry through this idea formally, I didn't see how to do it. I didn't see how to do it and therefore I can't say I was actively looking for it. I don't think so.

And then came a paper by Samet,[5] whom I have already mentioned. That's where I got this idea that we have a syntax, we have a set of statements, a list of statements; this list has to be complete and consistent. By complete we mean that every sentence – either the sentence itself or its negation – is in the list. Consistent means there are no logical contradictions in the list. A beautiful idea. You take a language and you say everything that can be said in that language. Everything. And then you look at lists, complete and consistent in this sense. And you say: that list *is* a state of the world. It's a beautiful idea. And this came out of Samet's paper.

I took that, developed it and formalised it. In some respects it was already formal. The fruition of this was the paper which was published in the *International Journal of Game Theory* in 1999, 'Interactive Epistemology',[6] which takes that and actually builds a semantic space out of this 'dictionary'. I was at Yale University for six weeks in '89, where I gave a course on interactive epistemology, and presented that idea. It took another ten years to publish it because I'm very slow to publish. Other stuff is going on and there's family and grandchildren, all kinds of stuff happening. It took ten years, but eventually I did publish, basically, the notes from the six weeks in '89 at Yale on interactive epistemology. So

---

[4] *Econometrica*, 55(1), pp. 1-18, 1987.

[5] D. Samet. Ignoring Ignorance and Agreeing to Disagree. *Journal of Economic Theory*, 52, pp. 190-207, 1990.

[6] *International Journal of Game Theory*, 28(3), pp. 263-300, 1999

that's how that came about. I'm convinced now that semantic game theory is fine, but the syntactic is more basic. Say it in English! If you have something to say, say it in English.

O.R.:
What is the relevance of epistemic logic? To what kind of topic is it best geared to contribute?

R.A.:
It's for exploring the foundations of game theory. You want to ask yourself: why do we look only at the objective mixed equilibria? Then one thing leads to another, you get the agreement theorem and then you have to ask yourself where this comes from. It's foundational work. It also develops a toolkit that enables you to think precisely about these things. It allows you to formulate precise questions.

Let me give you another application. I published this paper in 1995, 'Backward Induction and Common Knowledge of Rationality'.[7] People raised their eyebrows about it, and said: this is open to question, because according to Aumann's definition of rationality, people also have to behave rationally at nodes of the game tree that they themselves have excluded by previous actions. So they know they're not going to get there, but they still have to behave rationally. I wrote a lengthy discussion in that paper and I do think it makes sense and I think people should read the discussion part very, very carefully. Part of what ails the world is that people only look superficially at a paper. I really thought carefully about these problems. And I think the criticism is *not* justified.

But I always like to look at the other side of the coin. I said, OK, people don't like this idea that you have to behave rationally at nodes of the game tree that you yourself have excluded. They call these 'counterfactual conditionals'. In order to avoid using counterfactual conditionals, Adam Brandenburger and I replaced the idea of *knowledge* of something by *belief*, by the idea of *strong* belief. This means I continue to believe that everybody is rational until it becomes *logically* impossible to do so, until it's obvious we've reached a node of the game tree where somebody had to behave irrationally. And then we developed this idea of common strong belief of rationality. It seemed intuitive that this should lead to backward induction. Adam and I worked on this for five or six years and we couldn't prove it.

---

[7] *Games and Economic Behavior*, 8(1), pp. 6-19, 1995.

Then one day a paper crossed my desk written by Battigalli and Siniscalchi.[8] They did exactly that. Fantastic! It is a wonderful piece of work; it's very deep and it's very roundabout, but they did it! It's very good.

Battigalli and Siniscalchi did it using a very complex *semantic* model. In a semantic model it's difficult to express what it means for something to be logically impossible. In a semantic model, something that is logically impossible is represented by an empty set; but if an event is represented by an empty set, that doesn't mean it's logically impossible. You have to have a *universal* semantic model to say that, and these models are large and clumsy.

Now I have a student who is just completing his doctorate. His name is Itai Arieli, a very bright and deep young man. He managed to overcome this difficulty with a syntactic proof. 'Logically impossible' has a very transparent meaning in a syntactic model. Itai has a straightforward – well, straightforward may not be the right word – but he has a largely syntactic proof of this which solves the problem without encountering the conceptual difficulties of the semantic approach.

How does this relate to your question? The syntactic and the semantic models provide a *toolkit* for thinking about these problems. You want to fix some system, you need the screwdrivers. This is the toolkit for dealing with these problems and for attacking them, and thinking about them logically.

O.R.:
So if I understand correctly, you see this work in the characterisation of solution concepts as foundational for game theory. Could you think of it in the same way as the work in the foundations of decision theory or subjective probability, for example?

R.A.:
I think maybe it's less important in one-person decision theory. It does not play much of a role there. Savage,[9] Anscombe-Aumann,[10] it's not *essential*, I don't see applications there.

But let me say it's not just foundational stuff. I wrote a paper with Jacques Dreze[11] about rational expectations in games.

---

[8] P. Battigalli and M. Siniscalchi, Strong Belief and Forward Induction Reasoning. *Journal of Economic Theory*, 106(2), pp. 356-391, 2002.

[9] L. J. Savage, *The Foundations of Statistics*, New York: John Wiley, 1954.

[10] F. J. Anscombe and R. J. Aumann. A Definition of Subjective Probability. *The Annals of Mathematical Statistics*, 34(1), pp. 199-205, 1963.

[11] R. J. Aumann and J. H. Dreze, Rational Expectations in Games, *Amer-*

In some sense this grew out of the epistemic approach, which as I say started with the paper in the JME about subjectivity and correlation. This is more or less the latest development there; not the latest development in foundational terms but in *practical* terms.

In this paper with Jacques what we're really trying to promote is a revolution. We haven't been too successful yet. We're saying: Listen! Equilibrium is not the way to look at games. Nash equilibrium is *king* in game theory. Absolutely king. We say: No! Nash equilibrium is an interesting concept, it's an important concept, but it's not the most basic concept. The most basic concept should be: *to maximise your utility given your information*, in a game just like in any other situation. What does that imply when you make strong assumptions like common priors and common knowledge of rationality? Does it imply Nash equilibrium? No, it does not! It implies something else, and we can write formulas for it. That's the way to go.

In some sense the epistemic world of ideas led to this. This work started with a remark Jacques made. Jacques visited Israel in '96, and we went on a trip to Petra in Jordan with a whole group of economists who were visiting Israel at that time. I was sitting next to Jacques in the bus and he said to me: 'Bob, how would you define values in games that are not zero-sum?' We started thinking about it and a year or two later I went to Louvain-la-Neuve. We thought about it some more and eventually it grew into this paper. So in one sense the *question* didn't come from epistemic roots, but the *answer* came from epistemic roots. In two-person zero-sum games, common knowledge of rationality and common priors imply the value. That's it. You don't need the equilibrium concept. This gives a solid foundation to the value of two-person zero-sum games, which was missing before. Then you take that and you extend it to other games. I think that should replace Nash equilibrium as the fundamental concept of game theory. But it hasn't yet...

So epistemic considerations also lead to conclusions *outside* their own domain, which is the test of any important scientific theory or development. It's got to lead some place outside, not just inside.

O.R.:
In this paper with Dreze you make a strong case for analysing interaction in terms of common priors and common knowledge of rationality. Do you think these are fundamental assumptions?

R.A.:

Yes, I do. Let me put it this way: common priors is in my opinion a very reasonable assumption, *more* reasonable than common knowledge of rationality. People usually take it the other way around. They think that common knowledge of rationality is something they can buy; it's a strong assumption but it's something they can buy. But common priors? They ask 'Where did these come from?' *I* think common priors are more reasonable, because in some very basic sense you want somehow to account for differences in probability assessments. Why do people differ in their probability assessments? It seems reasonable that at *some* level it's because they had different inputs. So they got to know different things, and this is what accounts for the different posteriors. When you say that differences in probability assessments are due to different information, that is the common prior assumption. So this is really something very basic and very important.

Common knowledge of rationality is not that kind of thing at all. It's a very far-fetched assumption. I don't even know if *I'm* rational, but let's say I am. I'm willing to assume you're rational, but do I *know* that you *know* that I'm rational? It's not that common knowledge of rationality is something basic; it's some sort of *benchmark*.

The way we put it in the paper is: tic-tac-toe is a draw. What does that mean, that it *is* a draw? What it means is that under common knowledge of rationality the result will be a draw. It doesn't mean that a smart kid would not lose the game, because a smart kid might play for a win, or try, because that kid does not assume the other one knows that he is rational and so might play for some kind of trap and might lose as a result. Common knowledge of rationality is not something about the real world, it's a sort of a benchmark. It says: this is the way things work out when there's no irrationality in the system. If there is no irrationality in the system at all, this is how things work out. Everything else is calculated as a *departure* from the benchmark. It's like a no-friction assumption in physical systems, or the no-air-resistance assumption. When you have the formula for acceleration of a body under gravity, you let something fall, it assumes no air resistance. But there *is* air resistance. Does that make the formula for no air resistance unimportant? No, it's very important; you start out with the assumption that under *ideal* conditions, you have no air resistance, and then from there you can go and calculate what happens *with* air resistance. I don't think common knowledge of rationality is at all a reasonable assumption about the real world. I believe common priors *is* a reasonable assumption about the real world. Common priors is much more basic than CKR. But if you don't buy CKR, you won't get the value of a two-person zero-sum game.

So in order to answer Jacques' question, we do need CKR because he wants to generalize the value. The value of the game is the outcome when people play *correctly*. So when people play correctly, you have CKR in some sense. That's the *definition* of correct play.

O.R.:
You mention in that paper that the notion of value was something that was more or less forgotten but it returns when you look at a game from the perspective of common priors and common knowledge of rationality. Do you think there are other topics, especially in epistemic logic, that have been somewhat relegated to the background and should be given more prominence?

R.A.:
One should probably try to look at relaxing these conditions of CKR or of common priors. Shmuel Zamir spoke with me about 'approximately common priors'. One could try to look at rational expectations in specific classes of games. I can't think at the moment about something really foundational but my research, as far as I've experienced, never worked that way. Something seizes you and then you go in that direction. If you don't, then it hasn't seized you. Some kind of approximation to these ideal conditions is one question that can be asked. It's important to keep the application in mind, keep your purpose in mind. But to point to something specific, no, I don't know.

O.R.:
You mentioned the common knowledge of rationality assumption as a benchmark. Would you think the same of the assumption or the fact that agents in most epistemic logic systems are so-called logically omniscient?

R.A.:
Logical omniscience, that is a very important problem. It's a problem I've thought about a little. Not that I've gotten anywhere with it. I really don't know what to do with it. It is a very important problem, which does need treatment, but I don't know what to do with it.

O.R.:
Do you think recent work on unawareness could...

R.A.:
Unawareness, yes, those two: logical omniscience and unawareness, those are the two big open problems as far as I know. I think there's been work on them recently, though, but I haven't studied it. *Mea culpa.* I cannot venture an opinion as to whether the recent work really gets to the foundations of the problem. But those *are* the two outstanding problems. Maybe somebody has solved them in the meantime. Unawareness and logical omniscience: even though I am not omniscient – I do not claim to be omniscient – I'm unaware of satisfactory solutions to those two problems. But they are two very important problems, which do cry out for satisfactory treatment.

In that connection maybe I should mention something else which is maybe not particularly logical, but it has to do with logic. That is the problem of how much to calculate when you're optimising something. Most work that I know of, all the work I know of, does not take the cost of calculation into account. So the question is, how much do you calculate? If you really want to optimise the solution to something, how much should you calculate in order to find that solution? You want to get the cheapest possible solution without knowing what to expect. In order to even calculate the *time* it would take to optimise, you have to do a calculation, and that calculation in itself may not be worthwhile. When you're playing chess and the clock is ticking, how much do you calculate? It's not at all clear. Deep Blue had to solve this problem somehow. I don't know how they solved it. This is a very important problem, which has not been solved satisfactorily. It's in the same league as unawareness and logical omniscience.

O.R.:
So you say that the big issue or the big challenges for us now are about relaxing the fundamental assumptions?

R.A.:
No, I think logical omniscience and unawareness and how much to calculate, those are the three big outstanding issues at this time. The matter of relaxing, I don't think it's *that* important. It's not that basic. Those other things are more basic and I think they'll be more difficult to achieve. Relaxing is also relaxing the focus, in some sense: you see things more blurred. I like focus.

O.R.:
Are you optimistic about progress on these three questions?

R.A.:
Who knows? *Que sera, sera.* Those are Hilbert's three problems for epistemic game theory or logic. The third question is not really epistemic but it's also there and it has philosophical sides to it. But the unawareness and the logical omniscience, absolutely. Whether I'm optimistic, I don't know. It will be very nice if somebody solves them.

4
Johan van Benthem

Professor of Logic
University of Amsterdam & Stanford University
The Netherlands & USA

---

1. Why were you initially drawn to epistemic logic?

Initially, I was not 'drawn' to epistemic logic at all, but repelled by it. Like many of my generation of students around 1970, I considered epistemic logic an industrial variation on basic modal logic, the thing one can do for any sort of intensional operator in natural language, without any special redeeming interest. In particular, it was always a bit unclear whether epistemic logic had anything deep to say about the philosophical notion of knowledge, its professed motivation. And there was no technical compensation either to offset this lack of philosophical impact. No interesting mathematical questions arose from taking the epistemic focus: significantly, the first wave of technical research in the foundations of modal logic in the 1970s, where I was an eager young participant, bypassed it completely. This critical attitude is still common with some colleagues, especially when they go off the record. But for me, the negatives have eroded steadily, for reasons that I will explain, and by now epistemic logic has become a major interest of mine – one of those quiet but intense passions that gentlemen of advancing years are still capable of.

Let me first mention some external influences driving my change of heart and mind. The first was the TARK initiative around 1980, when it became clear that the multi-agent character of epistemic logic was a good fit with the then upcoming post-Turing challenge of understanding distributed computation by many processors, where the sparse abstraction level of a modal operator language seemed just right for bringing to light basic reasoning patterns about information and communication. Such surprising turns are of course frequent in intellectual history: a bad model for one phenomenon can turn out to be a very good model for another, often in ways completely unintended by the originators. In particular, multi-agent aspects were absent from the original epistemological motivation, while they are now crucial to the success of epistemic logic. Indeed, in the hands of Joe Halpern and his colleagues, TARK has made epistemic logic and computation in a broad sense into a viable and respectable endeavour, vital and still expanding. This program was enhanced still further by the emergence of common knowledge as a fundamental notion, which started a new strand, the logical study of knowledge of groups as collective entities. Thus epistemic logic acquired operators far beyond the basic modal language, linking it to dynamic logics of actions, and generating new mathematical questions after all, some of them unresolved until today. As an aside, this same impulse that brought me closer to epistemic logic was the point of farewell for many philosophers, who felt that the subject was falling in with bad company (the greasy hands of programmers and engineers), straying far from what should be and remain its true agenda of Knowledge in the purest sense. This fateful parting of the ways is still with us, but more on that below.

Epistemic logic became even more noticeable to me in the mid 1980s, with the emergence of 'auto-epistemic logic' that dealt with deep issues in nonmonotonic reasoning, and the reflective equilibria underpinning it, as analyzed by Bob Stalnaker, someone who of course has also lifted epistemic logic to greater philosophical heights generally in his work through the years. Around the same time, Bob Moore wrote a famous essay on knowledge and action showing how combinations of epistemic and dynamic logic generated interesting perspicuous base logics of information-driven action, where one could pose fundamental questions about knowing what plans do, while getting the right information to achieve one's goals. This, too, was a combination of philosophical ideas with computational ones, be it in Artificial Intelligence, a buffer state between philosophy and computer science where philosophers did venture at the time. At least, they did in the early years of Stanford's Center for the Study of Language and Information, when disciplinary barriers went down for a while, and it seemed that the intellectual world was being redefined for good.

The third and final influence that made epistemic logic a reality in my thinking was when logicians like me started learning about the epistemic foundations of Game Theory, pioneered by Robert Aumann in the 1970s. I came to see that epistemic logic, even if limited in itself, created exciting creative mixtures when added to other fields. That is also how I view the role of logic generally, as a catalyst for broader intellectual developments – or a spice, if you prefer cuisine. Moreover, this game-theoretic work changed my view of the position of logic in the university. In addition to the traditional allies of philosophy and mathematics, and in the second half of the 20th century, linguistics and computer science, logic had now also reached out successfully to economics and the social sciences, a new alignment to me.

But respect is not yet love. Epistemic logic became vitally important to me personally only when I started developing my own 'Logical Dynamics' program around 1990, as a study of the informational dynamics that drives human agency – and it has become increasingly central to it over the years. But let me tell that story a bit later.

2. What example(s) from your work, or work of others, illustrates the relevance of epistemic logic?

First of all, I do not think of epistemic logic as primarily a study of the philosophical, or even the psychological notion of knowledge. I would still submit that it has important things to say about knowledge, but that focus is not the best way of viewing what it really achieves. For a long time now, I have come to see epistemic logic as a theory of *semantic information*, in the sense of Carnap and Bar-Hillel, and indeed in the sense of most of the exact sciences that deal with semantic models. Of course, the models used are extremely simple (more on that below), but that is precisely why they allow us to see essential structures, and develop major themes in a perspicuous manner. To be sure, traditional epistemic logic is just a starting point here. Information comes in many kinds, from 'hard' to 'soft', and logic should, and can help us chart its basic varieties.

But epistemic logic only comes into its own when we make a further decisive step that was totally absent from the original development of the subject. This step, and a major interest of mine over the past decade is *information dynamics*, the explicit logical study of actions, events, and long-term processes that change information states of agents. This 'Dynamic Turn' has been my major interest over the last 20 years, witness my books *Language in Action* from 1991 (Elsevier Amsterdam & MIT Press Cambridge Mass.) and *Exploring Logical Dynamics* of 1996 (CSLI Publications Stanford). These books put the actions at centre stage that drive the logical and linguistic behaviour of single agents, such as acts of inference, interpretation, and assertion. This move transcends the usual boundaries of logic, since on this view, an act of inference has no privileged position as a way of acquiring knowledge: an observation may do just as well (direct perception may even be more reliable than a mathematical proof, witness the role of experiment in the physical sciences), and so does communication from a trusted source. Thus, Logical Dynamics takes a broad view of sources of knowledge: the logic serves to describe them, but logical inference is not the only acceptable producer of knowledge.

And even beyond this, by now, I have come to see the heart of rational agency as lying in conversation, argumentation, and other forms of interaction between several agents, and the version of epistemic logic that really powers this is not the traditional zoo of epistemic logics with their 'capitalist' names (S4, S5, KD45). It is rather the much richer framework of *dynamic epistemic logic*, a simple yet powerful paradigm for studying information flow by observation and communication, that has emerged over the past 15 years. This is the shape in which epistemic logic becomes broadly applicable to a wide range of phenomena, throwing new light beyond the usual static analysis of knowledge and other cognitive attitudes. Dynamic epistemic logics define informational events, allowing us to talk explicitly about what creates and modifies knowledge, and they have brought to light valid axioms of this information flow that are every bit as intriguing as the traditional epistemic axioms of closure or introspection – perhaps even more so. Even this is still at the heart of epistemology, though, and one of my main philosophical claims in the past ('Epistemic Logic and Epistemology: the State of their Affairs', *Philosophical Studies* 128, 2006, 49–76) has been that we can only understand the robustness of knowledge in the setting of the natural dynamic processes that deal with it. Or in my favourite computer science terms, 'no representation without transformation'.

But knowledge and information flow cannot be the whole story of rational cognition, and they may not even be the most important part of it. To me, the most striking aspect of rationality is not the statics of 'correctness' or being right, but the dynamics of *correction*, i.e., the quality of what we do when we revise beliefs that have been refuted. Knowledge may be the pure gold of the philosopher, but belief is the currency that really makes the whole cognitive enterprise run. It is beliefs that guide our actions, from taking the train to choosing avenues for new mathematical research. And the highest quality of rationality is not following some standard that automatically takes us from knowledge to knowledge, but rather the ability to jump to new beliefs, and correct these as we keep learning. Thus, Hintikka was right in studying knowledge and belief together, but their interplay becomes even more crucial when we consider the dynamics: forming beliefs and revising them is the art of living dangerously, a much greater cognitive skill than slavish correctness. Fortunately, it has just been found that the dynamics of *belief revision* and learning through self-correction can be studied very well in the dynamic-epistemic style, making them wholly comparable to the dynamics of information flow ('Dynamic Logic of Belief Revision', *Journal of Applied Non-Classical Logics* 17:2, 2007, 129–155). Incidentally, doing this right also involves a much richer view of information, ranging from 'hard' to 'soft' varieties, and corresponding kinds of information-driven revision actions. (Incidentally, I wonder what would have happened in the study of belief revision in the 1980s, now a separate industry, if this sort of dynamic logical apparatus had been around, rather than staid static 'doxastic logic'.)

My views on the importance of correction versus correctness are becoming quite radical. In the same light, I would now consider the traditional foundational search for consistency proofs of mathematical theories as misguided, since an essence of mathematical activity is the ability to construct new theories when old ones have gone bad. It should really be a relief to us all that Hilbert's Program failed for 'solving the foundational questions once and for all'. This is not to say that revision of mathematical theories is a well-understood phenomenon in my framework, and indeed, the dynamic-epistemic paradigm is still struggling with finding the best model for dealing with pure acts of reasoning.

My new monograph *Logical Dynamics of Information and Interaction* (Cambridge University Press, 2010) develops all this further into a logical study of information-driven agency, showing how epistemic logic in the dynamic sense is alive and thriving. I mention a few more themes from that book that represent major steps in extending the agenda of epistemic logic. One is incorporating more syntactic viewpoints on *fine-grained sorts of information*, and associated with this, acts of inference or introspection that make us explicitly aware of knowledge, beliefs, or even just of issues. A truly dynamic epistemic logic can deal with both syntactic and semantic information, and with acts that manipulate these – overcoming the traditional dichotomy between semantic and proof-theoretic paradigms for studying knowledge. In particular, one more crucial action comes to centre stage then: *questions* that raise issues directing future inquiry. It has been claimed that the whole history of science could be written more informatively as one of successive questions, rather than answers. Even without going that far, a true dynamic epistemic logic should be able to deal with questions as well as answers. Another essential aspect in my current view is the role of *long-term informational processes* such as learning procedures, or information-driven strategic behaviour in games. The initial dynamic step from static instantaneous knowledge to single events that change information was not enough. We also need an account of the longer term, as learning theorists have long seen – and again, modern epistemic logic can deliver such connections. I have even claimed in recent work ('The Information in Intuitionistic Logic', *Synthese* 167:2, 2009, 251–270) that this generates a further notion of information that is sui generis and crucial to an epistemology that cares for a richer view of agents: 'procedural information' that determines what single information update steps 'mean', making sense of them in a longer history of events.

There may even be one more traditional border that cannot be maintained in all this. We all know that true communication and even the scientific search for truth is driven by *values*. Everything we say and do has a colouring of evaluation, and tied up with this, preferences and goals for action. While this mixture is often seen as a sort of Platonic impurity, I would rather see it as essential to truly rational behaviour. The rational mind is in balance between information and evaluation. Purely informational interaction fails if we do not build a sort of group resonance where engaging in a shared activity is a value per se. Can we keep this evaluative aspect distinct from the informational one? I am beginning to doubt it, and indeed, it is of interest to see that techniques developed for dynamic epistemic logics are now crossing over into the study of preference, preference change, and deontic logics of changing obligations and norms ('For Better of for Worse: Dynamic Logics of Preference', in T. Grüne-Yanoff & S-O Hansson, eds., 2009, *Preference Change*, Springer, Dordrecht, 57 – 84). Again, these studies involve a delicate interplay of semantic and syntactic features, since evaluation may depend on both content and presentation.

Some critics may find this expansion of epistemology toward goal-driven agency a case of unprincipled disregard for established distinctions in philosophy. I myself would rather say that the program of knowledge in action outlined here represents a typical contribution that (dynamic) logic has to offer to philosophy, bringing to light common patterns in what are often considered disjoint fields of study.

The dynamic study of agency is not confined to philosophy, however. It applies equally well to (and draws on) the thriving areas of epistemic analysis of *agent systems* in computer science (the modern face of intelligent computation), and the epistemic-doxastic foundations of *game theory*. These examples also amply demonstrate what epistemic logic has to offer in its modern guise. Few areas of philosophical logic have had a similar impact.

3. What is the proper role of epistemic logic in relation to other disciplines, for instance mainstream epistemology, game theory, computer sciences or linguistics?

As I have said in earlier books in this series, I dislike this question, because 'proper role' smacks of unwarranted essentialism to me. But as I have noticed with successful Dutch colleagues in radio interviews, one should ignore the questions one dislikes, and just say what one wanted to say anyway when given the slightest chance. So, to me, epistemic logic is a general *logical information theory*, including both static information structure, the dynamic processes transforming this, and the changing attitudes of agents involved in all this. In particular, I think that the notion of information cannot be usefully analyzed without a matching logic of informational acts and processes, and hence what should be confronted with other disciplines is dynamic epistemic logic of one sort or another. Moreover, in this dynamic perspective, epistemic logic feeds naturally into a general account of information-driven *rational agency*, both in the common sense and in science. Thus, the border line with formal epistemology becomes fluid, but so does that with the philosophy of science.

Pursued in this new way, epistemic logic addresses common themes across philosophy, computer science, game theory, and many other fields, including linguistics. Indeed, logic plays two roles here. One is as a messenger: a medium of communication allowing for flow of ideas between fields in jointly understood terms. The other is as a perspective that can help revitalize existing discussions. Take the vexed issue of the definition of 'knowledge' in the philosophers' sense. As I have said, I myself think that its surplus over true belief is dynamic, having to do with robustness across many dynamic activities where information flows, by single agents, but especially also, in social settings of communication – the arena of Plato's *Dialogues* where the definition of knowledge first became an issue.

But is this merely 'interdisciplinarity' in the sense of aimless unpredictable wanderings? I have argued elsewhere that intellectual history shows continuities that only look discontinuous when one imposes the grid of standard disciplines ('Logic in Philosophy', in Dale Jacquette, ed., 2006, *Handbook of the Philosophy of Logic*, Elsevier, Amsterdam, 65–99). Epistemic logic itself is an example. Viewed in one way, it is a curiously segmented movie with disjoint parts in philosophy, computer science, and game theory. But if one thinks in terms of natural development of themes, a very different story emerges, of a continuous trend toward exploiting the potential of the semantic study of information in a multi-agent setting. I see it as one of the great missed opportunities in recent decades that philosophers have not followed these and other congenial developments beyond their borders, depriving themselves of the many inspiring conceptual insights that have come up in computer science, game theory, and other fields. Instead I often see defensive reactions, trying to explain why all this sort of work is 'shallow', 'irrelevant' and so on, at the same time when parts of analytical philosophy have reached heights of incrowd Scholasticism and elegant obscurity that not even the much-maligned Continental philosophers ever managed to achieve.

Against this background, some positive initiatives shine all the more. I mentioned the TARK conferences where congenial practitioners of many disciplines have met since the early 1980s, and one could cite more such events, such as the LOFT conferences, the recent Network on Rationality and Decision, or the new LORI conferences on Logic and Rational Interaction that have started in China. My impression is that a younger generation of logicians, epistemologists, and philosophers of science is forming these days that thinks much less in terms of traditional divisions, and that is open to external influences from other disciplines.

But now back to the role of epistemic logic in all this. Let me conclude with a caveat. One can have my positive view of epistemic logic as a public benefit between the disciplines mentioned in this question without claiming exclusive rights. One need not even claim exclusive rights as a flag-bearer for logic. For instance, one topic that intrigues me more and more is the dual existence of two major approaches to information in logical theory and its applications. Epistemic logic is an *explicit account* of information, adding new operators for knowledge to classical languages that are taken for granted. Likewise, dynamic epistemic logic is a conservative extension of classical logics with explicit operators for informational activities. By contrast, say, intuitionistic logic is an *implicit account* of information, changing the semantics of former languages so as to incorporate informational aspects, and thereby generating 'non-standard logics' with different sets of validities. This contrast is all-pervasive. For instance, Amsterdam-style 'dynamic semantics' changes the meaning of natural language semantics to incorporate actions that change information, issues, or what have you. By contrast, an explicit approach would keep the old truth-conditional meanings as before, but add a superstructure of dynamic logic for pragmatic acts of language use. The two souls can live within one breast. Even Hintikka himself goes the implicit way when developing his 'game-theoretic semantics' for expressions in natural language, where information about moves is crucial but remains implicit – whereas I have proposed an explicit epistemic game logic version in terms of what players know, using Hintikka's old ideas to clarify his new ideas ('The Epistemic Logic of IF Games', in R. Auxier & L. Hahn, eds., 2006 *The Philosophy of Jaakko Hintikka*, Schilpp Series, Open Court Publishers, Chicago, 481–513). To some, the implicit approach is deeper, since it seems to affect the very foundations of logic, sending unpredictable complex ripples through the whole discipline, whereas the explicit approach is just one of steady expansion of coverage. To me, both are natural stances, and their relationship a major issue in the philosophy of logic.

4. Which topics and/or contributions should have had more attention in late 20th century epistemic logic?

Question 4 is just the emotional version of Question 5. I will not deign it with a direct answer. But on the analogy of the preceding question, let me state another view of mine that seems relevant. It is a commonplace to say that epistemic logic leaves out various important features of real knowledge, cognition, and agency. While many people seem to be impressed by such criticisms, I find them thoroughly facile and predictable. Pointing out that Reality is more complex than some proposed formal model is easy, anyone can do it. Finding extreme simplifications that still leave us with important insights is hard, and only open to the most creative minds. Think of a conference where you are listening to a lecture. If you do not have the time or inclination to really think about what is being said, you can always raise a question of coverage and omitted aspects at the end: no deep thought required, and it never fails to impress the audience. But asking a truly perceptive question, on the other hand, requires real thought, and real immersion in the system that has been proposed. My sympathies are squarely with the small system builders and real engagement with what they do. Only when we understand the fruitful simplifications, I would go for disciplined extension toward the broader functioning of information and cognition, the way I have explained in the above.

And also, please recall my earlier point about the virtues of poverty. Simplified models may lead to surprising new applications and research directions, far beyond their initial territory and the intentions of their originators.

5. What are the most important open problems in epistemic logic and what are the prospects for progress?

I mentioned quite a few issues already in my answers so far. Dynamic epistemic logic is a research program in full swing that will generate new questions for quite a while to come. For instance, much remains to be understood about the logical interplay of various sorts of information (hard versus soft, semantic versus syntactic), and the many attitudes beyond knowledge and belief that agents can have in tune with these. There is every reason to assume that dynamic epistemic logic needs a much richer theory of agency than we have right now. Also, I feel that we are only beginning with the serious logical study of how information-driven knowledge works intertwined with revision, self-correction, and *learning* – and perhaps, if you go along with my earlier statements, even with the dynamics of *evaluation*. I have already indicated that these developments will probably lead to stronger connections between epistemic logic with the theory of agency in computer science, but also with learning theory in philosophy (as pioneered by Kelly and Hendricks), and between epistemic logic and game theory, where more attractive new models for rational agency may come about by endowing games with some epistemic agent structure. We seem to be moving toward an epistemic 'Theory of Play' rather than mere theory of games.

But more will have to happen, and the time seems ripe. I have already identified the mysterious interplay of external and internal views of knowledge of information, as exemplified by epistemic logic and intuitionistic logic. We need to understand this duality much better, and I am sure that we will. One further urgent desideratum is reintegrating basic themes from traditional philosophical logic, in particular *predication and quantification* over objects, merging knowledge of propositions with knowledge of objects. And once on this path from knowledge of propositions to that of objects, I would go even one step further. Think of learning and teaching. Most often, the precise propositional knowledge that we impart is only a means towards a more important end: knowledge of *methods*, acquiring skills that can work under different circumstances. You truly show that you know something when you can apply it, not in the same setting as the original, but in different ones. That is where we should go eventually.

At a higher abstraction level, I foresee several new alignments between fields triggered by modern developments in epistemic logic. I have already indicated some natural confluences inside philosophy between epistemic logic and learning theory, or between epistemic logic and logical theories of information. And I have pointed at the flourishing links between epistemic logic with computer science and game theory that keep inspiring 'joint ventures'. But there is more. A clear next goal down the line is forging links with *probability theory*, the major alternative paradigm in science and philosophy for analyzing information and agency. Probability crops up naturally and frequently even inside epistemic logic, once you start thinking about degrees of belief, global memory structure about the cognitive past, or mixed strategies for multi-agent behaviour in societies. In particular, there is no need to force a choice between 'competing paradigms' here: we should rather understand and cultivate the interface. And finally, I foresee new alliances between epistemic logic and cognitive science, looking at the intricacies of social cognition and intelligent interaction far beyond the usual studies of single-agent inference that have dominated the interface so far.

To conclude, please recall that in my view, all these strands are not disjoint from philosophy, as they form a natural intellectual whole, including attractive agenda extensions for traditional epistemology and related fields. So let me return to the beginning of this interview. Epistemic logic started as a tool for analyzing the philosophical notion of knowledge and its standard problems. It may not have done too good a job at that. But what it has to offer these days is an enticing view of what epistemology *could become*, provided that philosophers are willing to loosen up, and engage in the same sort of agenda dynamics that has taken epistemic logic to its current vitality and scope.

5

Giacomo Bonanno

Professor
University of California, Davis, USA

---

1. Why were you initially drawn to epistemic logic?

In my case it took a long and tortuous path to reach the pasture of epistemic logic. My academic career started with a degree in Law from the university of Turin, Italy. During my (compulsory) military service, following graduation, I learned of a scholarship to pursue graduate studies in economics in England. I was lucky enough to obtain it and went to Cambridge University for a one-year Master's degree in economics. I then had the privilege of spending one year at the Mathematics Institute of the University of Warwick, doing research (in catastrophe theory) under the supervision of Christopher Zeeman. The next three years were spent in the Ph. D. program at the London School of Economics, under the supervision of Oliver Hart and John Sutton. My thesis was in theoretical Industrial Organization, which is essentially game theory applied to issues of interaction among firms. I continued to do research in Industrial Organization during the next two years as a Research Fellow at Nuffield College, Oxford. During that time, however, I became somewhat dissatisfied with the field of Industrial Organization since there was a widespread impression that, by a suitable choice of game and solution concept, almost any kind of behavior could be rationalized. In some cases the difference in results could be explained in terms of different structural assumptions about the industry (e.g. differentiated products versus homogeneous products), while in other cases the divergence of results was due to a different sequencing of moves in the game and/or the use of different solution concepts. That was the time of the "refinement of Nash equilibrium" program, which yielded as many as thirty different proposals for "the rational solution" of a non-cooperative game! I thus became interested in more foundational issues and, in particular, in pure game theory and the conceptual underpinnings of the various solution concepts proposed in the literature. While I was at Oxford I had a long and illuminating conversation with Michael Bacharach, who encouraged me to pursue my new research interests. In 1987 I moved to the University of California, Davis. In 1994, while on sabbatical leave at Harvard university, I attended a course in modal logic in the philosophy department, taught by Charles Parsons. I clearly saw the relevance of modal logic in general, and epistemic logic in particular, to game theory. This cemented the new direction in my research, centered on the application of various branches of modal logic (epistemic, doxastic, temporal, conditional logic) to the analysis of games. While pursuing this line of inquiry I became aware of related research in other fields, notably artificial intelligence, philosophy, logic and cognitive psychology. In 1996 Mamoru Kaneko and Philippe Mongin invited me to co-organize the second LOFT (Logic and the Foundations of Game and Decision Theory) conference. I found the interdisciplinary character of the conference to be very stimulating and was happy to remain involved - together with the computer scientist Wiebe van der Hoek - in the organization of this biannual event, which is now in its ninth edition. The LOFT conferences brought to light the fact that the tools and methodology that were used in the investigation of foundational issues in game theory were closely related to those already used in other fields, notably computer science, philosophy and logic. Epistemic logic turned out to be the common language that made it possible to bring together different professional communities. It became clear that insights gained and the tools used in one field could benefit researchers in a different field. Indeed, new and active areas of research have sprung from the interdisciplinary exposure provided by the LOFT conferences.[1] I myself have benefited greatly from entering into contact with researchers from other disciplines, in particular Johan van Benthem and his collaborators and students.

---

[1] See: http://www.econ.ucdavis.edu/faculty/bonanno/loft.html. There is now a substantial overlap between the LOFT community and the community of researchers who are active in another biannual event, namely the TARK conferences (Theoretical Aspects of Rationality and Knowledge: www.tark.org).

2. What example(s) from your work, or work of others, illustrates the relevance of epistemic logic?

My interest in epistemic logic originated in game theory and thus I will limit my observations to the relevance of epistemic logic to game theory.

Game theory can be thought of as being composed of two separate modules. The first module consists of a formal language for the description of interactive situations, that is, situations where several individuals take actions that affect each other. This language provides alternative descriptions, from the more detailed one of extensive forms to the more condensed notions of strategic form and coalitional form. The language of game theory has proved to be useful in such diverse fields as economics, political science, military science, evolutionary biology, computer science, mathematical logic, experimental psychology, sociology and social philosophy. The unifying role of the game-theoretic language has been a major achievement in itself. The second module is represented by the collection of solution concepts. A solution concept associates with every game in a given class an outcome or set of outcomes. Most of the debate in game theory has centered on this module, in particular on the rationale for, and interpretation of, different solution concepts. The "epistemic foundation program" in non-cooperative game theory tries to determine what assumptions on the beliefs and reasoning of the players are implicit in various solution concepts. The first step in this direction was taken by Bernheim (1984) and Pearce (1984) whose aim was to identify, for every game, the strategies that might be chosen by rational and intelligent players who know the structure of the game and the preferences of their opponents and who recognize each other's rationality and reasoning abilities. In order to address this issue, one needs to answer two questions: (1) under what circumstances is a player rational? and (2) what does 'mutual recognition' of rationality mean? While there is agreement among game theorists about some essential ingredients of the notion of rationality (the most basic ingredient being: choosing a strategy which is optimal given one's beliefs), it seems that there is no clear and commonly accepted definition of rationality in general.[2] On the other hand, there does seem to be agreement that the notion of

---

[2] Witness, for example, the debate between Robert Aumann and Ken Binmore concerning backward induction in perfect-information games [Aumann (1996), Binmore (1996)].

'mutual recognition' of rationality is to be interpreted as 'common belief' of rationality. This is the point of entry for epistemic logic: how does one formalize the notion of common belief? What are the properties of common belief? Does common *knowledge* of rationality have different implications from common *belief* of rationality? etc. While Bernheim and Pearce captured the notion of common belief of rationality only informally, the most important contributions from the point of view of establishing a connection with epistemic logic were (implicitly) Aumann (1987) and (explicitly) Stalnaker (1994, 1996). Both authors follow a semantic approach, using structures that had been introduced in philosophy and logic in the early 1960s by Kripke (1963).[3] Aumann restricts attention to the notion of knowledge, while Stalnaker allows for the more general notion of belief. The connections between epistemic logic and the game-theoretic literature on the epistemic foundations of non-cooperative solution concepts are reviewed in detail in Battigalli and Bonanno (1999).

Other branches of modal logic have also proved to be useful in elucidating game-theoretic concepts. For instance, extensive-form games have a clear connection to the branching-time frames studied in temporal logic and thus temporal logic can shed light on solution concepts for extensive-form games [Bonanno (2001, 2002)] or on the conceptual content of properties such as perfect recall [Bonanno (2004)].

Logicians [notably, Johan van Benthem (2001, 2010)] and computer scientists [notably, Joe Halpern (2001, 2002)], have recently been very active in the study of game-theoretic concepts from the standpoint of modal logic. For a recent survey of the literature see van der Hoek and Pauly (2006).

3. What is the proper role of epistemic logic in relation to other disciplines, for instance mainstream epistemology, game theory, computer sciences or linguistics?

I once heard Robert Aumann remark at a workshop in Stanford that one important difference between game theory and pure mathematics is that game theory is about life, everyday life. I would argue that the same is true of epistemology, broadly conceived. We make decisions (some more important, some less so)

---

[3] A *syntactic* approach that uses the language and methods of epistemic logic to analyze games with ordinal payoffs can be found in Bonanno (2008).

on a daily basis, and we reach those decisions after having formed, consciously or unconsciously, a mental representation of the situation we face. A central part of this mental representation consists of our beliefs about the world and about the likely consequences of alternative courses of action. In a social context we also try to put ourselves in the shoes of other people in an attempt to anticipate their choices and their potential reactions to our actions. Sometimes our beliefs turn out to be correct and sometimes they are revealed to be based on misinformation or an incorrect appraisal of the situation, in which case we are prompted to revise our initial beliefs. I would argue that mental states, and in particular beliefs, are the essential element of any theory of social interaction. This is particularly true in game theory. There certainly are settings, such as a game of chess or a game of bridge, where one can refer to "the game" being played, with a common recognition among the players of the rules of the game and of the objectives of each player. However, such settings are rare. Most of the time we interact socially in situations where there are no rigid rules to be followed. In such cases Ann's mental representation of the situation might be different in some (perhaps small, but essential) elements from Bob's mental representation. "The game" being played according to Ann might be quite different from "the game" perceived by Bob, and yet each player may be certain that "the game" is "evident" or "commonly understood". Is it possible for Ann to believe that it is common belief between her and Bob that they are playing game G_1, while at the same time Bob believes that it is common belief between him and Ann that they are playing game G_2, when G_1 is a different game from G_2? In order to answer this type of questions one needs to turn to logic, in particular, epistemic logic. It turns out that such a situation can indeed arise; however, if we replace 'belief' with 'knowledge' then it cannot arise.

Given the importance of beliefs in everyday life, and social interaction in particular, it is important to have at our disposal tools that enable us to reason about beliefs, whether they are beliefs about facts or, perhaps more importantly, beliefs about the beliefs of other people. This is what I see as the important role of epistemic logic in relation to game theory. The use of epistemic logic can bring to light subtle, yet important, points. For example, the standard tool in game theory for representing interactive epistemic states is the notion of information partition [see, for example, Aumann (1987)] which embodies the assumption of correct belief, also called *knowledge*. It turns out that in this setting the notion of *common* knowledge displays the same properties as the notion of individual knowledge; in particular, if something is *not* common knowledge then this fact itself is common knowledge. Does the same hold in the case of (possibly incorrect) beliefs? If one postulates the strongest rationality properties for individual beliefs,[4] are those properties also satisfied for *common* belief? The answer turns out to be negative [Bonanno and Nehring (2000a)], with implications for the notion of a common prior [Bonanno and Nehring (1999)], which is central to the theory of games of incomplete information. In an interactive context where individuals' mental states are characterized in terms of both knowledge (those propositions of which an individual is absolutely certain) and belief (those propositions about which the individual is confident) then the notion of intersubjective consistency becomes even more subtle [Bonanno and Nehring (2000b)]. It is hard to see how one could reason about interactive mental states without the tool of epistemic logic.

4. Which topics and/or contributions should have had more attention in late 20th century epistemic logic?

It is hard for me to answer this question, since my interest in epistemic logic (and modal logic in general) was motivated by potential applications to game and decision theory. Hence my knowledge of the epistemic logic literature is somewhat limited. I am probably myself unaware of contributions to epistemic logic that might be potentially important from the point of view of game theory. I certainly believe that if game theorists had been aware, earlier on, of epistemic logic and the pioneering work of Kripke (1963), Hintikka (1962) and Lewis (1969), the literature on the epistemic foundations of non-cooperative game theory would have followed a simpler path and the theory of games of incomplete information [Harsanyi (1967-68)] might have been developed in a simpler, more transparent and more straightforward way.[5]

---

[4] That is, *consistency* (if you believe A then you do not also believe not-A), *positive introspection* (if you believe A then you believe that you believe A) and *negative introspection* (if you do not believe A then you believe that you do not believe A).

[5] Here I am echoing Robert Aumann's remark [Aumann (1999, p. 295)] that the infinite-hierarchies-of-beliefs construction is very convoluted.

5. What are the most important open problems in epistemic logic and what are the prospects for progress?

Recently I have become interested in the topic of belief revision: how should a "rational" individual incorporate new information into her existing body of beliefs? Belief revision is important in game theory, especially in dynamic games with imperfect information. If a player finds herself at an information set that was ruled out by her prior beliefs (she attached zero probability to that event), she needs to revise those beliefs (in particular, her beliefs about past moves of her opponents) by incorporating the new information. Bayes' Rule is not applicable to updating after zero probability events. Does "rationality" impose any constraints on how one should update those beliefs?

The theory of belief revision has been developed mainly by philosophers and computer scientists. The dominant approach is known as the AGM theory, following the pioneering contribution of Alchourrón, Gärdenfors and Makinson (1985). The AGM theory deals with the transition from a belief state to a new belief state in response to a piece of information. Information is treated as veridical and the "success axiom" is assumed, which requires that information be believed. While belief revision is an active area of research, there are important open problems that ought to be addressed or further explored.

The first problem concerns the notion of information. Belief revision is about incorporating reliable information into one's beliefs. What constitutes reliable information? Years ago, perhaps, a photograph could be taken as "indisputable evidence". Nowadays, with the advent of sophisticated image-editing software, photographs can be manipulated to misrepresent facts or to create the appearance of an event that did not happen. For example, in March 2004 a political advertisement for George W. Bush, as he was running for president, showed a sea of soldiers at a public event; later the Bush campaign acknowledged that the photo had been doctored, by copying and pasting several soldiers.[6] Videos and voice recordings are, nowadays, equally manipulable. What can one trust as a source of reliable information? The testimony of a witness? A newspaper article? A book? A television news report? A claim by the president of the USA? Many of us rely on the internet for information. Can material found on the internet

---

[6] For an interesting account of photo tampering throughout history, see http://www.cs.dartmouth.edu/farid/research/digitaltampering/.

be trusted as accurate? In Footnote 6 I gave a reference to a web page reporting photo tampering throughout history: can one be sure that the information given there is correct? Note that I am not necessarily implying that incorrect information is conveyed maliciously with the intent to mislead, although sometimes this does in fact happen.[7] It may simply be the case that an initial piece of incorrect information gets reproduced (in good faith) by different sources and thus becomes "confirmed" information. One is left wondering if, nowadays, there is *any* source of information that is completely reliable. The theory of belief revision needs to address the issue of belief formation and revision in a world where no information can be fully trusted. Furthermore, in a social context, the incentives to convey wrong information need to be studied and incorporated into a theory of belief revision.

A second problem is how to deal with sequences of items of information which are in partial or full contradiction with each other. This can happen when the same source, over time, provides contradicting information or when different sources provide conflicting information (e.g. different experts). To some extent, this issue has been studied in the literature on *iterated* belief revision, where various principles have been suggested (for example, the principle that the most recent item of information should prevail over earlier ones). However, the proposed principles seem rather *ad hoc* and in need of a firmer foundation.

A third problem concerns the notion of "minimal" belief change. The AGM theory is often referred to as a theory incorporating the principle that beliefs should be changed in a minimal way, so as to ensure that there is minimal loss of prior beliefs. While this is true

---

[7]Even reputable sources can sometimes be manipulated. A clear illustration of this can be found in the following newspaper report (The Sacramento Bee, September 1, 2000): "Mark J. made a big bet in mid-August that Emulex shares would decline [...] Instead they soared, leaving him with a paper loss of almost $100,000 in just a week. So J. took matters into his own hands. [...] On the evening of August 24, he sent a fake press release by e-mail to Internet Wire, a Los Angeles service where he had previously worked, warning that Emulex's chief executive had resigned and its earnings were overstated. The next morning, just as financial markets opened, Internet Wire distributed the damaging release to news organizations and Web sites. An hour later, shareholders in Emulex were $2.5 billion poorer. And J. would soon be $240,000 richer. [...] The hoax [...] was revealed within an hour of the first news report and Emulex stock recovered the same day. Still, investors who [believing the fake news release] panicked and sold their shares, or had sell orders automatically executed at present prices, are unlikely to recover their losses"

when new information is compatible with prior beliefs, in the case where the new information contradicts the earlier beliefs, there is really no constraint imposed by the AGM postulates in terms of preserving as many of the old beliefs as possible. Indeed one way of revising beliefs, which is consistent with the AGM postulates, if to form a new belief set consisting exclusively of the learned information and anything that can be logically deduced from it. More work needs to be done on what minimal belief change entails.

A fourth problem is related to the observation that, while epistemic logic deals with how people *should* reason, how they *should* form beliefs and how they *should* respond to new information, an important area of research concerns how people *actually* reason and process information [Holyoak and Morrison (2005), Johnson-Laird (2006)]. It seems to me that much would be gained by attempting to integrate insights from both areas of research.

Finally, while beliefs are arguably the most important mental states, there are other mental states that deserve a similar in-depth investigation, in particular intention, prediction and emotions and their interactions with belief formation and belief change.

Although some contributions have addressed the "open" problems listed above, more work needs to be done in order to gain a better understanding of the issues involved. Furthermore, interdisciplinary approaches might be useful in shedding light on connections between seemingly different research questions.[8]

References

Alchourrón, Carlos, Peter Gärdenfors and David Makinson (1985), On the logic of theory change: partial meet contraction and revision functions, *The Journal of Symbolic Logic*, 50: 510-530.

Aumann, Robert (1987), Correlated equilibrium as an expression of Bayesian rationality, *Econometrica*, 55: 1-18.

Aumann, Robert (1996). Reply to Binmore. *Games and Economic Behavior*, 17: 138–146.

Aumann, Robert (1999), Interactive epistemology I: Knowledge, *International Journal of Game Theory*, 28: 263-300.

---

[8] For example, it turns out that the mathematical structures used in decision theory to model revealed preference can be re-interpreted in terms of partial belief revision in accordance to the AGM postulates [Bonanno (2009)].

Battigalli, Pierpaolo and Giacomo Bonanno (1999), Recent results on belief, knowledge and the epistemic foundations of game theory, *Research in Economics*, 53: 149-225.

Bernheim, Douglas (1984), Rationalizable strategic behavior, *Econometrica*, 52: 1007-28.

van Benthem, Johan (2001), Games in dynamic epistemic logic, *Bulletin of Economic Research*, 53: 219-248.

van Benthem, Johan (2010), *Logic in games*, Amsterdam University Press, Texts in Logic and Games, Amsterdam (to appear).

Binmore, Ken (1996). A note on backward induction. *Games and Economic Behavior*, 17: 135-137.

Bonanno, Giacomo (2001), Branching time logic, perfect information games and backward induction, *Games and Economic Behavior*, 36 : 57-73.

Bonanno, Giacomo (2002), Modal logic and game theory: two alternative approaches, *Risk Decision and Policy*, 7: 309-324.

Bonanno, Giacomo (2004), Memory and perfect recall in extensive games, *Games and Economic Behavior*, 47: 237-256.

Bonanno, Giacomo (2008), A syntactic approach to rationality in games with ordinal payoffs, in: Giacomo Bonanno, Wiebe van der Hoek and Michael Wooldridge (editors), *Logic and the Foundations of Game and Decision Theory*, Texts in Logic and Games Series, Amsterdam University Press, 59-86.

Bonanno, Giacomo (2009), Rational choice and AGM belief revision, *Artificial Intelligence*, 173: 1194-1203.

Bonanno, Giacomo and Klaus Nehring (1999), How to make sense of the common prior assumption under incomplete information, *International Journal of Game Theory*, 28: 409-434.

Bonanno, Giacomo and Klaus Nehring (2000*a*), Common belief with the logic of individual belief, *Mathematical Logic Quarterly*, 46: 49-52.

Bonanno, Giacomo and Klaus Nehring (2000*b*), Intersubjective consistency of Knowledge and Belief in: Martina Faller, Stefan Kaufmann and Marc Pauly (editors), *Formalizing the dynamics of information*, CSLI Publications, Stanford, 27-50.

Halpern, Joseph (2001), Substantive rationality and backward induction, *Games and Economic Behavior*, 37: 425-435.

Halpern, Joseph (2002), Characterizing the common prior assumption, *Journal of Economic Theory*, 106: 316-355.

Harsanyi, John (1967-68), Games with incomplete information played by "Bayesian players", Parts I-III, *Management Science*, 8: 159-182, 320-334, 486-502.

Hintikka, Jaakko (1962), *Knowledge and belief*, Cornell University Press, 1962.

van der Hoek, Wiebe and Marc Pauly (2006), Modal logic for games and information, in: Johan van Benthem, Patrick Blackburn, and Frank Wolter (editors), *Handbook of Modal Logic*, Elsevier, 1077-1148.

Holyoak, Keith J. and Robert G. Morrison (editors) (2005), *The Cambridge handbook of thinking and reasoning*, Cambridge University Press.

Johnson-Laird, Philip (2006), *How we reason*, Oxford University Press.

Kripke, Samuel (1963), A semantical analysis of modal logic I: normal propositional calculi, *Zeitschrift für Mathematische Logik und Grundlagen der Mathematik*, 9: 67-96.

Lewis, David (1969), *Convention: a philosophical study*, Harvard University Press.

Pearce, David (1984), Rationalizable strategic behavior and the problem of perfection, *Econometrica*, 52: 1029-50.

Stalnaker, Robert (1994), On the evaluation of solution concepts, *Theory and Decision*, 37: 49-74.

Stalnaker, Robert (1996), Knowledge, belief and counterfactual reasoning in games, *Economics and Philosophy*, 12: 133-163.

6

Adam Brandenburger

J.P. Valles Professor

Stern School of Business, New York University, USA

Origins of Epistemic Game Theory[1]

6.1 Introduction

In a 1935 article titled "Perfect Foresight and Economic Equilibrium," Oskar Morgenstern wrote about how members of a social system form expectations, expectations about expectations, and the like:

> "[T]here is exhibited an endless chain of reciprocally conjectural reactions and counter-reactions.... The remedy would lie in analogous employment of the so-called Russell theory of types in logistics. This would mean that on the basis of the assumed knowledge by the economic subjects of theoretical tenets of Type I, there can be formulated higher propositions of the theory; thus, at least, of Type II. On the basis of information about tenets of Type II, propositions of Type III, at least, may be set up, etc." [14, 1935, pp.174-176]

Morgenstern had written earlier [13, 1928, p.98] about a "battle of wits" between Sherlock Holmes and Professor Moriarty in which each tries to think about what the other is thinking, about what

---

[1] I am indebted to Sergei Artemov, Amanda Friedenberg, Willemien Kets, Alex Peysakhovich, Ariel Ropek, Gus Stuart, and Natalya Vinokurova for important input and to Harold Kuhn and Robert Leonard for kindly providing copies of Wolfe [22, 1955]. Financial support from the Stern School of Business is gratefully acknowledged.

the other is thinking he (the first) is thinking, and so on. In 1935, Morgenstern added the suggestion that such ingredients could be captured via formal mathematical-logical methods.

It took approximately fifty years until a formalization of Morgenstern's bold but more-or-less forgotten idea re-appeared–in the modern subfield of game theory called epistemic game theory.

What accounts for the delay?

6.2 The Protective (Maximin) Criterion

I suggest that the answer, at least in part, is that von Neumann, the intellectual giant with whom Morgenstern embarked on the systematic construction of game theory, put different considerations center-stage.

A key theme in von Neumann's famous 1928 paper [20, 1928] is what payoff an individual player–or group of players–can be guaranteed to get. In adopting a maximin strategy, a player "is protected against his adversary 'finding him out'" [20, 1928, p.23]. This is von Neumann's theory of play in two-player zero-sum games. It is also his theory for n-player zero-sum and general-sum games. For these games, von Neumann introduces the characteristic function of a game, defined by assigning to each subset of players the total payoff that subset can guarantee itself via coordinated choice of actions–regardless of what coordinated actions the players outside the subset might choose. In brief, each subset is assigned the maximin payoff of that subset.

The maximin criterion obviates Morgenstern's epistemic considerations. Each player (or set of players) considers the 'worst-case' scenario in terms of which strategies the other player (or players) might choose. No attempt is made to arrive at a prediction about what other players will do by putting oneself in their shoes and thinking about what they might be thinking, and so on. In *Theory of Games and Economic Behavior* [21, 1944], von Neumann and Morgenstern made this very explicit: "Nor are our results for one player based upon any belief in the rational conduct of the other" [21, 1944, p.160].

It seems that von Neumann's agenda dominated Morgenstern's.

One more observation from *Theory of Games and Economic Behavior*: Does the game model determine how a game is played? Von Neumann and Morgenstern said no:

"[W]e shall in most cases observe a multiplicity of solutions. Considering what we have said about interpreting solutions as stable 'standards of behavior' this has a simple and not unreasonable meaning, namely that given the same physical background different 'established orders of society' or 'accepted standards of behavior' can be built...." [21, 1944, p.42]

Alternatively put, the reason for multiplicity is that outcomes are under-determined by the game model. Additional factors–of a more 'intangible' kind–also matter. In this way, the von Neumann-Morgenstern philosophy can be thought of as a kind of indeterminism.

6.3 The Equilibrium Criterion

Nash's reformulation of the n-player theory removes both the cooperative and the maximin aspects of the von Neumann-Morgenstern theory. He puts the question of what is rational individual play firmly on the table. We might even say "... back on the table," since we could argue that this question is closer to what Morgenstern was talking about in 1935. But, we will see a big difference in how the two saw the question.

Nash wrote:

"We proceed by investigating the question: what would be a 'rational' prediction of the behavior to be expected of rational[ly] playing the game in question? By using the principles that a rational prediction should be unique, that the players should be able to deduce and make use of it, and that such knowledge on the part of each player of what to expect the others to do should not lead him to act out of conformity with the prediction, one is led to the concept of a solution defined before." [16, 1950]

The key components of Nash's argument are that: (i) associated with each game is a unique correct way to analyze that game; (ii) this way is accessible to the players themselves; (iii) each player makes the best choice of strategy for him/herself.

In (ii), Nash is saying that a player can step outside the game, so to speak, and adopt the role of observer or analyst. If players do this, they see the game exactly the way an observer does. The player vs. observer issue is very interesting—and we will come back to it.

Right now, we focus on (i). The equilibrium selection program recognizes the importance of this step for Nash, and tries to narrow down the set of equilibria for any given game to a single point. (The position can be taken that without a theory of equilibrium selection, Nash's argument breaks down. As an aside, we note that Hillas and Kohlberg [10, 2002, Section 8.2] distinguish the refinements program from the selection program. They define the former as concerned with identifying necessary—not sufficient—conditions on candidates for the 'right' equilibrium of a game.)

In any case, Nash supposes uniqueness and (via step (ii)) concludes that each player will reach a correct conclusion about what strategies the other players choose. Add rationality (step (iii)), and we arrive at Nash equilibrium.

Nash's uniqueness assumption is very different from the multiplicity that von Neumann and Morgenstern saw as a natural and desirable state of affairs.

Also, Nash's conclusion that each player is correct about the others' choices is very different from how Morgenstern—who wrote about "faulty, heterogeneous foresight" [14, 1935, p.174]—appears to have been thinking.

It is well known that von Neumann did not receive Nash's idea positively. (See Shubik [19, 1992, p.155].) Was von Neumann convinced later? It seems not. Multiplicity remained central to von Neumann's picture of game theory. Here is a report of what he said at a 1955 conference on game theory at Princeton:

> "The discussion opened with a statement by von Neumann in justification of the enormous variety of solutions which may obtain for n-person games. He pointed out that this was not surprising in view of the correspondingly enormous variety of observed stable social structures; many differing conventions can endure, existing today for no better reason than that they were here yesterday." [22, 1955, p.25]

However, with the rise of the Nash equilibrium concept, von Neumann-Morgenstern multiplicity—which can also be called von Neumann-Morgenstern indeterminism—was pushed aside.

6.4 Types

Harsanyi [9, 1967-8] wanted to analyze uncertainty about the structure of a game–specifically, about the players' payoff functions. To this end, he introduced the fundamental concept of a player's "type," which, he asserted, could be used to encode what the player believes the payoff functions to be, what the player believes other players believe the payoff functions to be, and so on indefinitely. (This brilliant idea was given a formal justification much later–by Armbruster and Böge [1, 1979], Böge and Eisele [5, 1979], Mertens and Zamir [12, 1985], and others.)

For a finite set X, write $\mathcal{M}(X)$ for the set of probability measures on X. Given sets X^i, for $i = 1, \ldots, n$, let $X = \times_i X^i$, and, for each i, $X^{-i} = \times_{j \neq i} X^j$. Now fix, for each player i, a finite set S^i of strategies for player i. Harsanyi's formalism consists of, for each i,

- a finite set T^i of types for player i;

- a map $f^i : T^i \to \mathcal{M}(T^{-i})$;

- a map $g^i : T^i \to S^i$;

- a map $h^i : S \times T \to \mathbb{R}$ (the reals).

(Sometimes, the map g^i is to $\mathcal{M}(S^i)$ rather than S^i. But, we can 'purify' these maps by defining new type spaces $U^i = T^i \times [0, 1]$. Of course, this entails extending the framework to infinite type spaces. We omit the details in this short piece.)

Example 1 Figures 4.1 and 4.2 are a simple illustration (taken from Myerson [15, 1985, p.241]) of Harsanyi's formalism. Start with Figure 4.1. Ann has one possible type t^a. Bob has two possible types t^b and v^b. Ann's type t^a assigns probability 3/5 to t^b and probability 2/5 to v^b. Also, type t^a plays U and has payoff function h^a. Of course, both types t^b and v^b for Bob assign probability 1 to t^a. Type t^b plays L and has payoff function h^b. Type v^b plays R and has payoff function $\tilde{h}^b$. The payoff functions are depicted in Figure 4.2.

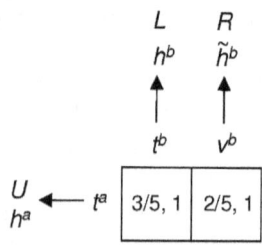

Figure 4.1

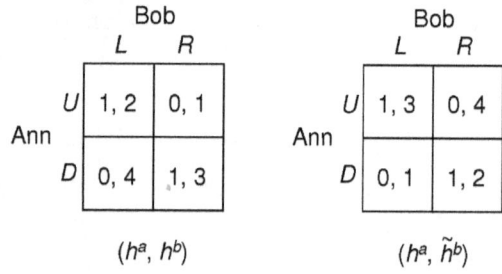

Figure 4.2

In this example, Ann has a simple induced hierarchy of beliefs about the payoff functions. She assigns probability 3/5 to Bob's having payoff function h^b and probability 2/5 to his having payoff function $\tilde{h}^b$. She assigns probability 1 to Bob's assigning probability 1 to her having the payoff function h^a. And so on to higher levels.

One more item from Harsanyi's formulation: He required each type to optimize as follows. (Given maps $\varphi^i : X^i \to Y^i$, let $\varphi^{-i} = \times_{j \neq i} \varphi^j$.) For each player i and each type t^i for i,

$$\sum_{t^{-i} \in T^{-i}} f^i(t^i)(t^{-i}) h^i(g^i(t^i), g^{-i}(t^{-i}), t) \geq \\ \sum_{t^{-i} \in T^{-i}} f^i(t^i)(t^{-i}) h^i(s^i, g^{-i}(t^{-i}), t) \qquad (6.1)$$

for all $s^i \in S^i$. This, of course, is the condition for Bayesian equilibrium. It is easy to check that Ann's type and both of Bob's types in Example 4.1 satisfy the condition. We will refer back to this item shortly.

6.5 Epistemic Game Theory

Here is another example that uses Harsanyi's formalism.

Example 2 There are two types t^a and v^a for Ann, and two types t^b and v^b for Bob. Figure 5.1 depicts the probabilities associated (via the maps f^i) with each type for either player. For example, type t^a assigns probability $1/2$ (resp. $1/2$) to Bob's being type t^b (resp. v^b). The diagram also depicts the strategy associated (via the maps g^i) with each type. There is only one payoff function for each player–namely, the payoff function depicted in Figure 5.2–and so we have suppressed the maps h^a and h^b.

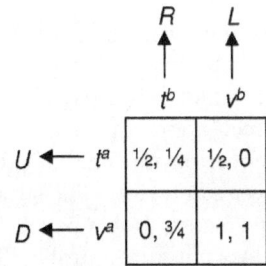

Figure 5.1

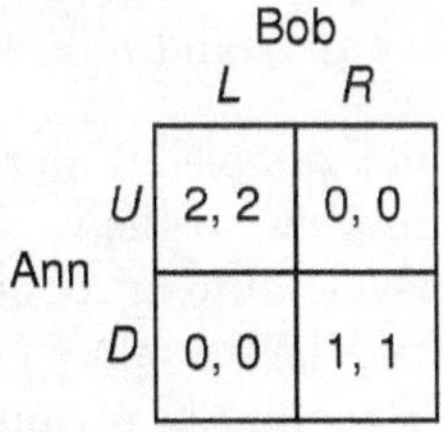

	Bob	
	L	R
Ann U	2, 2	0, 0
Ann D	0, 0	1, 1

Figure 5.2

Figure 5.1 is a simple example of the type structures used in epistemic game theory. Suppose the actual state of the world is (U, t^a, R, t^b). We can then calculate Ann's hierarchy of beliefs over the strategies chosen. We see that Ann assigns probability $1/2$ to Bob's choosing R and probability $1/2$ to Bob's choosing L. Ann also assigns: (i) probability $1/2$ to the event "Bob chooses R and assigns probability $1/4$ to her (Ann's) choosing U and probability $3/4$ to her choosing D"; and (ii) probability $1/2$ to the event "Bob chooses L and assigns probability 0 to her (Ann's) choosing U and probability 1 to her choosing D"; and so on to higher levels. We can likewise calculate Bob's hierarchy of beliefs over the strategies.

We can also talk about the "rationality" or "irrationality" of a type. The rationality criterion is exactly inequality (4.1) above. Thus, each player is rational: Type t^a for Ann optimally chooses U and type t^b for Bob optimally chooses R. We also see that Ann assigns probability 1/2 to Bob's being rational (type t^b) and probability 1/2 to Bob's being irrational (type v^b). Bob assigns probability 1/4 to Ann's being rational (type t^a) and probability 3/4 to Ann's being irrational (type v^a). Again, we can continue to higher levels.

There are two important differences between the two examples:

- In Example 4.1, we did not calculate the hierarchies of beliefs over the strategies. The reason is that, in this example, if Ann were to come to know Bob's payoff function, she would be certain–and correct–about the strategy he chooses. Her hierarchy of beliefs over payoff functions determines her hierarchy of beliefs over strategies. The situation in Example 5.1 is different. Ann knows Bob's payoff function (there is only one), but she does not assign probability 1 to his actual choice of strategy. (Likewise from Bob's perspective.)

- In Example 4.1, all types optimize–i.e., satisfy inequality (4.1). In Example 5.1, some types optimize and some do not. In particular, types v^a and v^b are irrational.

These two new features in Example 5.1 mark the transition from the world of Bayesian equilibrium to the world of epistemic game theory (EGT). The first new feature introduces into game theory the idea of uncertainty about the strategies in a game–in addition to uncertainty about the structure of a game. The second new feature is the introduction of irrationality, or belief in irrationality, or belief about belief in irrationality, and the like.

It is true that, in principle, both features are expressible in Harsanyi's formalism (more precisely, provided one does not insist on Bayesian equilibrium). But, in practice, this was not how his formalism came to be used. Indeed, in the numerical examples Harsanyi himself used in [9, 1967-8] to illustrate his framework, neither feature is present.

A clear break is evident in the papers by Bernheim [4, 1984] and Pearce [17, 1984]. Written during the height of the equilibrium refinements program, while many people were working on trying to narrow down the set of Nash equilibria in a game, these two papers challenged the view that Nash equilibrium was the inevitable starting point of analysis in the first place.

Rather than banish uncertainty about strategies (as Nash did), Bernheim and Pearce make this uncertainty central. (But, they did not treat irrationality.) Ann has subjectively formed probabilities about Bob's choice of strategy, constrained only by the assumption that she believes him to be rational, she believes he believes her to be rational, and so on. Call this the assumption of "common belief" of rationality. Actually, Bernheim and Pearce assumed "common knowledge" of rationality. On common knowledge, see Aumann [2, 1976] and Lewis [11, 1969], and also the remarkable earlier work by Friedell [8, 1967] (re-discovered by Barry O'Neill). The belief-knowledge distinction is very important in EGT, but we will not go into it here.

It is intuitively clear that the assumption of rationality and common belief of rationality implies that each player chooses an iteratively undominated strategy–i.e., a strategy that survives iterated elimination of strongly dominated strategies. EGT proper began with formal proofs of this assertion–and of an appropriate converse–using type structures like the one in Figure 5.1. This set the foundation for subsequent research in the area. Up to the present, EGT has studied the epistemics of irrationality as well as rationality, the epistemics of game trees, and the epistemics of weak dominance, among other issues.

Here is a general definition of the field today (for which I am grateful to Sergei Artemov): *EGT makes epistemic states of players an input of a game and devises solution concepts that take epistemics into account.*

6.6 Indeterminism Again

Where do the type structures of EGT come from? For a given game, what determines the appropriate type structure?

The answer is that the type structure is to be understood as part of the description of the situation being studied–on a par with the strategy sets, the set of outcomes, and the payoff functions (and the information sets in a tree). Remember that a player's payoffs are that player's own evaluation of the possible outcomes of the game. What a player believes, what a player believes other players believe, etc., is also subjective. Both payoffs and beliefs are subjective inputs into the game model. (In decision theory, Savage [18, 1954, p.3] uses the illuminating term "personalistic" in place of "subjective.") Neither payoffs nor beliefs can be deduced from other components of the model; they must both be described.

An epistemic analysis will, in most cases, depend on the particular type structure used. In such cases, the outcome of the analysis will be under-determined by the classical game model. We are back to the same situation as the one von Neumann and Morgenstern found–and for broadly similar reasons. There are 'intangible' as well as 'tangible' components of the model, and we should not expect determinacy from the second component alone.

This said, there are type structures that have a special status. These are type structures that, in one sense or another, contain all possible beliefs. Various notions of such a structure have been given–terminal structures (Böge and Eisele [5, 1979]), canonically-built structures (often called universal structures, see Mertens and Zamir [12, 1985]), and complete structures (Brandenburger [6, 2003]).

Epistemic analysis on such structures can yield sharp results. An example is the important paper by Battigalli and Siniscalchi [3, 2002], which states epistemic conditions on a complete structure and derives from them unique outcomes in a number of games of applied interest.

6.7 Epistemics and Logic

Most work in EGT has used the tools of Polish spaces, Borel probability measures, etc.

But the topic of large type structures turns out to be one where tools from mathematical logic have also proved useful. These tools have the virtue of being explicit about what methods of reasoning are being used to think about a game.

Taking a logic perspective on epistemic conditions of the "I believe that you believe that I believe..." kind might lead one to suspect that some type of self-reference could arise in game theory. Similar to occurrences of self-reference in other areas (most famously, of course, in set theory), could this lead to an impossibility result?

Jerry Keisler and I [7, 2006] investigated this question, and found the following theorem: Let the players use a language that includes first-order logic (and symbols for the relations in the type structure). Then a large such structure–formally, a structure which is complete relative to this language–does not exist.

An interpretation is that the type structure is a tool used by the analyst to describe the game. If this tool is also available to the players, then a difficulty can arise. Should the language(s) used be restricted to avoid the impossibility? Or, is the key to maintain a sharp distinction between the players and analyst of a game? Logical tools seem well-suited to investigating these (and other) questions in EGT.

6.8 Conclusion

Morgenstern's bold idea of using the tools of formal logic to talk about how members of a social system think, about how they think about what other members think, and so on, was far ahead of its time. But now, in the form of EGT, it has found a home.

6.9 REFERENCES

[1] Armbruster, W., and W. Böge, "Bayesian Game Theory," in Möschlin, O., and D. Pallaschke (eds.), *Game Theory and Related Topics*, North-Holland, Amsterdam, 1979.

[2] Aumann, R., "Agreeing to Disagree," *Annals of Statistics*, 4, 1976, 1236-1239.

[3] Battigalli, P., and M. Siniscalchi, "Strong Belief and Forward-Induction Reasoning," *Journal of Economic Theory*, 106, 2002, 356-391.

[4] Bernheim, D., "Rationalizable Strategic Behavior," *Econometrica*, 52, 1984, 1007-1028.

[5] Böge, W., and T. Eisele, "On Solutions of Bayesian Games," *International Journal of Game Theory*, 8, 1979, 193-215.

[6] Brandenburger, A., "On the Existence of a 'Complete' Possibility Structure," in Basili, M., N. Dimitri, and I. Gilboa (eds.), *Cognitive Processes and Economic Behavior*, Routledge, 2003, 30-34.

[7] Brandenburger, A., and H.J. Keisler, "An Impossibility Theorem on Beliefs in Games," *Studia Logica*, 84, 2006, 211-240.

[8] Friedell, M., "On the Structure of Shared Awareness," Paper #27, Center for Research on Social Organization, University of Michigan, 1967.

[9] Harsanyi, J., "Games with Incomplete Information Played by 'Bayesian' Players, I-III," *Management Science*, 14, 1967-8, 159-182, 320-334, 486-502.

[10] Hillas, J., and E. Kohlberg. "Conceptual Foundations of Strategic Equilibrium," in Aumann, R., and S. Hart (eds.), *Handbook of Game Theory*, Vol. III, Elsevier, 2002, 1597-1663.

[11] Lewis, D., *Convention: A Philosophical Study*, Harvard University Press 1969.

[12] Mertens, J-F., and S. Zamir, "Formulation of Bayesian Analysis for Games with Incomplete Information," *International Journal of Game Theory*, 14, 1985, 1-29.

[13] Morgenstern, O., *Wirtschaftsprognose, eine Untersuchung ihrer Voraussetzungen und Mödglichkeiten*, Springer Verlag, 1928.

[14] Morgenstern, O., "Vollkommene Voraussicht und wirtschaftliches Gleichgewicht," *Zeitshrift für Nationalökonomie*, 6, 1935, 337-357. Reprinted as "Perfect Foresight and Economic Equilibrium," in Schotter, A. (ed.), *Selected Economic Writings of Oskar Morgenstern*, New York University Press, 1976, 169-183.

[15] Myerson, R., "Bayesian Equilibrium and Incentive-Compatibility: An Introduction," in Hurwicz, L., D. Schmeidler, and H. Sonnenschein (eds.), *Social Goals and Social Organization*, Cambridge University Press, 1985, 229-260.

[16] Nash, J., "Non-Cooperative Games," doctoral dissertation, Princeton University, 1950.

[17] Pearce, D., "Rational Strategic Behavior and the Problem of Perfection," *Econometrica*, 52, 1984, 1029-1050.

[18] Savage, L., *Foundations of Statistics*, Wiley, 1954.

[19] Shubik, M., "Game Theory at Princeton, 1949-1955: A Personal Reminiscence," in Weintraub, E.R. (ed.), *Toward a History of Game Theory*, Duke University Press, 1992, 151-164.

[20] Von Neumann, J., "Zur Theorie der Gesellschaftsspiele," *Mathematische Annalen*, 100, 1928, 295-320.

[21] Von Neumann, J., and O. Morgenstern, *Theory of Games and Economic Behavior*, Princeton University Press, 1944.

[22] Wolfe, P. (ed.), Report of an informal conference on *Recent Developments in the Theory of Games*, Department of Mathematics (Logistics Research Project), Princeton University, January 31-February 1, 1955.

7
Hans van Ditmarsch

Researcher

University of Sevilla, Spain

---

1. Why were you initially drawn to epistemic logic?

I worked for a while at the Open University of the Netherland in course development. One product of this time was a course on logic, consisting of a Dutch-language textbook Logica voor informatica (Logic for computer science) [25] that I co-authored with Johan van Benthem and others. And this is how I met Johan. The book was accompanied by a number of syllabi, more or less books as well, that contained more technical details, exercises and answers, and case studies. The second of these syllabi would have a part on epistemic logic and Wiebe van der Hoek would write that part. And that is how I met Wiebe.

Now one of these Open University things is that apart from feeding students *content*, epistemic logical content in this case, you also give them a *case study*, something to apply that content to. Wiebe told me about this game I had never heard of, called Cluedo, that might constitute a suitable case study in epistemic logic... Cluedo is a boardgame wherein six players try to determine *who* killed a Mr. Black with *what* weapon and in *what* room, by asking other players questions about their ownership of playing cards for suspects, weapons, and rooms. I liked the idea, and it got into the tentative plans for that course. Things did not turn out that way at the time. Eventually we had too much course material, the material on epistemic logic was cut down a bit, and the case study on Cluedo never materialized.

The topic of Cluedo kept roaming my brain, and came out again when, several years later, I discussed PhD topics with possible supervisors. One of them, Johan van Benthem! And also my paths crossed with Wiebe again at that stage, as he became informally involved with the supervision of my PhD. Part of my PhD work was an analysis of the dynamics of knowledge in the game of Cluedo [27], the characterization of knowledge in models for card deals [32], and even some tentative observations on how to win Cluedo using that novel understanding of structure and dynamics [26]. The epistemic logic of knowledge games as Cluedo is now well understood, also in a temporal epistemic setting [7].

2. What example(s) from your work, or work of others, illustrates the relevance of epistemic logic?

How to win games of Cluedo does not illustrate the relevance of epistemic logic. It seems to much of a toy problem. A particular move in the game however resurfaces in more relevant problem areas. If in Cluedo you pass on your move to the next player, this implies implicitly that you do not yet know who the murderer is (what the murder cards are). For the next player, this may be so informative that (s)he can determine what the murder cards are. In other words, your inability to win is the informative source for another player to be able to win. This smells like a familiar phenomenon to any epistemologist. The so-called Moore-sentence 'p is true and you don't know that' becomes false after being announced [20, 13]; it is an *unsuccessful update* [10, 31]. The rise — if one may call it that — of dynamic epistemic logic as an independent specialism in logics of knowledge has certainly brought the analysis of such phenomena to the foreground of philosophical research. So much for relevance to the wider logical and philosophical community.

It is also relevant for technological applications. Security protocol analysis is full of examples where behaviour interpreted like the intention to keep a secret leaks information and therefore reveals the secret. Now one can make a link between the Moore-sentence like phenomena in epistemic logic and such information leaks. I championed a particular link in what is now known as the Russian Cards Problem [28]. (It was improperly named after the 2000 Moscow Mathematics Olympiad — but it is much older, and dates back to at least Thomas Kirkman [14].) Given a distribution of seven cards over three players, two players each holding three cards (sender and receiver) can communicate their hand of cards to each other by public communications without the third player (eavesdropper), who holds a single card only, learning from any card except his own who holds it. One can compare various protocols consisting of announcements that attempt to solve the problem on how effective they keep the eavesdropper ignorant. It turns out that there are announcements after which the eavesdropper is ignorant but such that the sender does not know that the eavesdropper is ignorant — from which the eavesdropper learns the cards of the sender. Worse, there are announcements after which the sender knows that the eavesdropper is ignorant, but such that the eavesdropper does not know that the sender knows that the eavesdropper is ignorant — and again the eavesdropper learns the sender's cards from that reflective observation. One has to take into account not just static knowledge about the card deal and the results of moves, but also dynamic knowledge about the protocol that the players are executing to get to know the cards of others. Knowledge of the sender's protocol is like knowledge of the sender's intentions, and that is very revealing for the eavesdropper: the sender may give away the secret because of her intention to keep it. This is exactly like a Moore-sentence becoming false because it is being announced. Epistemic languages with explicit protocol description are currently much in fashion [23, 35].

3. What is the proper role of epistemic logic in relation to other disciplines, for instance mainstream epistemology, game theory, computer sciences or linguistics?

I wouldn't know much about the *proper* role. But I know about the actual role in relation to other disciplines. And I am in favour of relations across disciplines, and everybody seems to be in favour of that. So what can I say...

I'd say that the the role of epistemic logic in relation to mainstream epistemology is surprisingly limited. What is almost all of dynamic epistemic logic about? It is almost all about expressive variations on the $S5$ notion of knowledge, that satisfies that known propositions are true, and that known and unknown propositions are known. Slightly further off, and still in the public eye, the slightly weaker $KD45$ notion of belief, that satisfies the latter but not the former (and that also satisfies consistency of beliefs). But that's it. Now this whole story of dynamics, preferences, graded beliefs, conditional belief and knowledge, temporal epistemics, probabilistic epistemics, and so on, could just as well apply – I think – to the much wider area of mainstream epistemology. (Most of the constructions and proofs apply to *any* modality, not just to $S5$ or $KD45$.) One such productive link is the analysis of Fitch's paradox of knowability in dynamic epistemic logic [9, 22, 4, 21]. I hope others may follow.

Currently emerging issues in dynamic epistemics involve expressive power and succinctness of logical languages [15, 19]. Both interest me greatly. The former seems to relate more to computer science and the latter to linguistics.

Let us take the issue of succinctness. As said, after announcing "p is true and you don't that," it becomes false. You now know p. And this is always the case after such an announcement. In the logic of public announcements, this is formalized by the expression $[p \wedge \neg Kp]\neg(p \wedge \neg Kp)$ that contains the public announcement operator $[p \wedge \neg Kp]$. It can be reduced by logical equivalences to an expression without such an operator. It is equivalent to $(p \wedge \neg Kp) \to \neg[p \wedge \neg Kp](p \wedge \neg Kp)$, which is equivalent to $(p \wedge \neg Kp) \to \neg([p \wedge \neg Kp]p \wedge [p \wedge \neg Kp]\neg Kp)$, which in its turn is equivalent, using several similar reduction steps, to $(p \wedge \neg Kp) \to \neg(((p \wedge \neg Kp) \to p) \wedge ((p \wedge \neg Kp) \to ((p \wedge \neg Kp) \to K((p \wedge \neg Kp) \to p)))$. The last is an expression without public announcement operators. It is longer than the formula we started out with. That one is more *succinct*. This notion of succinctness also applies to a logic, in a precise technical sense. In the case of the example we just saw, the last formula has also a shorter, equivalent, formulation in epistemic logic. But there are formulas in multi-agent public announcement logic where the *shortest* equivalent formulation in epistemic logic is a *lot* longer. (Where 'a lot' can be made precise in terms of complexity classes.) We therefore say that (not just this example but the whole) public announcement logic is more succinct than epistemic logic.

But, after all, for every formula with public announcements we can find an equivalent one without it. The logics are *equally expressive*. Other dynamic epistemic logics are more expressive than the basic epistemic logic, e.g., when we add common knowledge operators to epistemic logic, or (even more expressive than the previous) to public announcement logic. Expressiveness questions are relevant to computer scientists—in programming language design one faces the similar challenge to specify behaviour into 'words', i.e., some formal language.

Now for me the issue of succinctness is also related to the logical form of natural language, to linguistics therefore. Natural language can also be very succinct. A picture may say more than a thousand words, but twenty words may be enough to say something about a thousand pictures. One can be very clear and succinct in natural language. Logics that keep the logical forms of such sentences in the process of formalizing them may be similarly succinct.

Epistemic logic relates to many more disciplines... Under 'prospects for progress' I will address a relation to game theory. Epistemic logic relates to other modal logics — it is merely *one* such modal logic. There is currently a movement from dynamic epistemic logic to doing dynamics the same way for other modal logics, such as logics of preference, intention, obligation, permission. The *dynamics* of epistemic logic seem very relevant to model dynamics of other cognitive, intentional, and pragmatic phenomena [24, 11, 1, 17].

4. Which topics and/or contributions should have had more attention in late 20th century epistemic logic?

I find this a very hard question. I can only observe, going back 10 or 20 years from now, what I would at that time have expected now to have been done. For example, despite the general interest in models for restricted rationality, bounded reasoning, and more cognitively grounded formal logics, there still is not a standard logic of bounded reasoning. Certainly lots of good work are going on, I much like the approach by Alechina with various collaborators on bounded resources (such as [3]). But it seems to me that a common ground has not been reached. Possibly, a more psychologically motivated approach with probabilistic aspects and bounded ('working') memory might come to the fore at some stage.

Somewhat along that line, I find it surprising that the attention for alternatives for $S5$ in the 70s and 80s has ground to a halt. Wolfgang Lenzen at the time suggested a number of weaker alternatives to $S5$ [16]. Possibly, integrating weaker logics than $S5$ with logics for bounded reasoning at the same time may be a fruitful combination. Well, here we are already verging on the open problems for the future; and maybe $S5$ is simply it, for system building, despite the grumblings of epistemologists. Ah, almost — yet another topic that did not come off the ground as I, from a somewhat limited perspective, had expected, is multi-agent frame characterization. Anything that describes that agent a knows more than agent b, that agents a, b, c, d are in a certain hierarchy of knowledge where one knows more or less than the other depending on roles and permissions, all nicely tied in with correspondence theory. There is work as by Lomuscio on multi-agent characterization of interpreted systems with maximal uncertainty about the state of other agents [18]. Multi-agent correspondence theory is a personal interest. With my long-term collaborators Barteld Kooi and Wiebe van der Hoek I have a publication on characterizing *models*—not frames!—for card deals [32], and a totally unrelated recent publication on agents that know more than other agents in the actual state (i.e., locally), but not throughout the model [33].

5. What are the most important open problems in epistemic logic and what are the prospects for progress??

Here again, I can mainly observe what are the most important open problems in epistemic logic *for me*, and of course, on all of those the prospects for progress are very good. This is not because I'm such a great guy, but because I apparently refrain from becoming interested in problems for which I consider prospects for progress bad. A self-fulfilling prophecy! I can only hope that my intuitions about these prospects are accurate.

I became recently much interested in logics for knowledge *and awareness*. Combinations of knowledge and awareness have been around for a while, basically since an eminently readable detailed investigation by Fagin and Halpern in the late 1980s [8], and can roughly be distinguished in syntactic approaches, where the agent is aware of a limited number of formulas — and therefore not of their deductive closure, so this is a bounded-reasoning approach — and semantic approaches, where the vocabulary of the agent is limited but the agent is perfectly rational. These logics have the interest of economists [12]. I work on these matters with Tim French, where our ideas, initially proposed as a sideline in [30] involve incorporating bisimulation quantification and a suitable notion of bisimulation given unawareness.

A strangely open problem is how to model formally such a purely epistemic notion as *lying* (and related concepts like bluffing, deceiving, cheating, ...). I am not talking about liars' paradoxes and such, but the everyday standard sort of lying where you are informing others of something you know to be false, why you do this, how you get away with it, or not, etc. You'd say that this should have been done a long time ago. But not so. Baltag mentions the topic occasionally [5, 29], and interesting ongoing work is being done by Caminada [6] (containing various references to related work in epistemics that does not model lying as a dynamic notion changing the beliefs of agents), and there is a manuscript I co-authored [34].

Part of truths, lies, and knowledge is about playing the lying *game* well. How many lies can you tell and still get away with?

And if others do it too? And what sort of playing for information is possibly anyway? Traditionally, epistemic logicians have modelled the dynamics of knowledge from a descriptive perspective: if you say this, I will learn that, and if you say that, I will learn something else. But how about modelling knowledge dynamics from a prescriptive perspective? If I want you to learn (get to know) this or that, what should I say? If we both want to learn the truth about p ("where is the secret code to the one million dollar safe"), and we both want to get to know this before the other, and if, dreadfully, each of us has distinct information that needs to be combined before either of us can act, what sort of questions and answers and announcements should we make in order to achieve that goal? These matters, and how to obtain equilibria, are discussed in the manuscript [2].

7.1 REFERENCES

[1] T. Ågotnes. Action and knowledge in alternating-time temporal logic. *Synthese (Special Section on Knowledge, Rationality and Action)*, 149(2):377–409, 2006.

[2] T. Ågotnes and H. van Ditmarsch. What will they say? - public announcement games. *Synthese (Knowledge, Rationality & Action)*, To Appear.

[3] Natasha Alechina, Brian Logan, and Mark Whitsey. A complete and decidable logic for resource-bounded agents. In Nicholas R. Jennings, Charles Sierra, Liz Sonenberg, and Milind Tambe, editors, *Proceedings of the Third International Joint Conference on Autonomous Agents and Multi-Agent Systems (AAMAS 2004)*, pages 606–613, New York, July 2004. ACM Press.

[4] P. Balbiani, A. Baltag, H. van Ditmarsch, A. Herzig, T. Hoshi, and T. De Lima. 'Knowable' as 'known after an announcement'. *Review of Symbolic Logic*, 1(3):305–334, 2008.

[5] A. Baltag and S. Smets. Dynamic belief revision over multi-agent plausibility models. In G. Bonanno, W. van der Hoek, and M. Wooldridge, editors, *After-conference Proceedings of LOFT 2006*, Texts in Logic and Games. Amsterdam University Press, 2008.

[6] M.W.A. Caminada. Truth, lies and bullshit; distinguishing classes of dishonesty. Workshop Social Simulation at the Inter-

national Joint Conference on Artificial Intelligence (SSIJCAI), 2009.

[7] C. Dixon. Using temporal logics of knowledge for specification and verification–a case study. *Journal of Applied Logic*, 4(1):50–78, 2006.

[8] R. Fagin and J.Y. Halpern. Belief, awareness, and limited reasoning. *Artificial Intelligence*, 34(1):39–76, 1988.

[9] F.B. Fitch. A logical analysis of some value concepts. *The Journal of Symbolic Logic*, 28(2):135–142, 1963.

[10] J.D. Gerbrandy. *Bisimulations on Planet Kripke*. PhD thesis, University of Amsterdam, 1999. ILLC Dissertation Series DS-1999-01.

[11] D. Grossi and F.R. Velázquez-Quesada. *Twelve Angry Men*: A study on the fine-grain of announcements. In X. He, J.F. Horty, and E. Pacuit, editors, *Logic, Rationality, and Interaction. Proceedings of LORI 2009*, pages 147–160. Springer, 2009. LNCS 5834.

[12] A. Heifetz, M. Meier, and B.C. Schipper. Interactive unawareness. *Journal of Economic Theory*, 130:78–94, 2006.

[13] J. Hintikka. *Knowledge and Belief*. Cornell University Press, Ithaca, NY, 1962.

[14] T. Kirkman. On a problem in combinations. *Camb. and Dublin Math. J.*, 2:191–204, 1847.

[15] B.P. Kooi. Expressivity and completeness for public update logics via reduction axioms. *Journal of Applied Non-Classical Logics*, 17(2):231–254, 2007.

[16] W. Lenzen. Recent work in epistemic logic. *Acta Philosophica Fennica*, 30:1–219, 1978.

[17] F. Liu. *Changing for the Better: Preference Dynamics and Agent Diversity*. PhD thesis, University of Amsterdam, 2008. ILLC Dissertation Series DS-2008-02.

[18] A.R. Lomuscio. *Knowledge Sharing among Ideal Agents*. PhD thesis, University of Birmingham, Birmingham, UK, 1999.

[19] C. Lutz. Complexity and succinctness of public announcement logic. In *Proceedings of the Fifth International Joint Conference on Autonomous Agents and Multi-Agent Systems (AAMAS 06)*, pages 137–144, 2006.

[20] G.E. Moore. A reply to my critics. In P.A. Schilpp, editor, *The Philosophy of G.E. Moore*, pages 535–677. Northwestern University, Evanston IL, 1942. The Library of Living Philosophers (volume 4).

[21] J. Salerno, editor. *New Essays on the Knowability Paradox*. Oxford University Press, Oxford, UK, 2009.

[22] J. van Benthem. What one may come to know. *Analysis*, 64(2):95–105, 2004.

[23] J. van Benthem, J.D. Gerbrandy, T. Hoshi, and E. Pacuit. Merging frameworks for interaction. *Journal of Philosophical Logic*, 38:491–526, 2009.

[24] J. van Benthem and S. Minica. Toward a dynamic logic of questions. In X. He, J.F. Horty, and E. Pacuit, editors, *Logic, Rationality, and Interaction. Proceedings of LORI 2009*, pages 27–41. Springer, 2009. LNCS 5834.

[25] J. van Benthem, H. van Ditmarsch, J. Ketting, J.S. Lodder, and W.P.M. Meyer-Viol. *Logica voor informatica (Logic for Computer Science)*. Addison-Wesley, Amsterdam, 2003. Third revised edition.

[26] H. van Ditmarsch. The description of game actions in cluedo. In L.A. Petrosian and V.V. Mazalov, editors, *Game Theory and Applications*, volume 8, pages 1–28, Commack, NY, USA, 2002. Nova Science Publishers.

[27] H. van Ditmarsch. Descriptions of game actions. *Journal of Logic, Language and Information*, 11:349–365, 2002.

[28] H. van Ditmarsch. The russian cards problem. *Studia Logica*, 75:31–62, 2003.

[29] H. van Ditmarsch. Comments on 'the logic of conditional doxastic actions'. In K.R. Apt and R. van Rooij, editors, *New Perspectives on Games and Interaction*, Texts in Logic and Games 4, pages 33–44. Amsterdam University Press, 2008.

[30] H. van Ditmarsch and T. French. Simulation and information. In J. Broersen and J.-J. Meyer, editors, *Knowledge Representation for Agents and Multi-Agent Systems*, pages 51–65. Springer, 2009. LNAI 5605. Presented at LOFT 2008 and KRAMAS 2008.

[31] H. van Ditmarsch and B.P. Kooi. The secret of my success. *Synthese*, 151:201–232, 2006.

[32] H. van Ditmarsch, W. van der Hoek, and B.P. Kooi. Descriptions of game states. In G. Mints and R. Muskens, editors, *Logic, Games, and Constructive Sets*, pages 43–58, Stanford, 2003. CSLI Publications. CSLI Lecture Notes No. 161.

[33] H. van Ditmarsch, W. van der Hoek, and B.P. Kooi. Knowing more - from global to local correspondence. In C. Boutilier, editor, *Proceedings of 21st IJCAI 2009, Pasadena*, pages 955–960, 2009.

[34] H. van Ditmarsch, J. van Eijck, F. Sietsma, and Y. Wang. On the logic of lying. Under submission, 2008.

[35] Y. Wang, L. Kuppusamy, and J. van Eijck. Verifying epistemic protocols under common knowledge. In *TARK '09: Proceedings of the 12th Conference on Theoretical Aspects of Rationality and Knowledge*, pages 257–266, New York, NY, USA, 2009. ACM.

8
Melvin Fitting

Professor

CUNY Graduate Center, NYC, USA

---

1. Why were you initially drawn to epistemic logic?

Whys are often mysterious things. Certainly I was drawn to epistemic logic. Could it have been otherwise? Chance seems to have had a considerable role, but I don't really know. All I have is the way it happened.

Undergraduate college was an engineering college, Rensselaer Polytechnic Institute. My reasons for attending Rensselaer were straightforward: it was in my home town, and was the only college I knew much about. I was a mathematics major because I had a talent for the subject, a liking for it, and a drive to work on it. The college tried its best to give students a broad education. In particular, it had a good Philosophy department, and I took several of its courses, including a few somewhat pedantic logic courses. The chairman, who taught an ethics course I attended, understood that I had aptitude for logic and lent me Jaakko Hintikka's book, "Knowledge and Belief," which had appeared maybe 3 years earlier. As it happened, I didn't much like the book. (Fortunately, opinions can change with time.) My only previous exposure to logic had been axiomatic, while Hintikka's approach was semantic. His book contained an early version of semantic tableaus for modal logic, or at least this could be extracted from what was there, and I didn't like tableaus either, since I was only comfortable with axiomatics. At this point, if I had met me I would have erred considerably in any characterization of my deep aptitudes and interests.

For graduate school I attended Yeshiva University, starting in 1968. This was one of two universities whose PhD programs had logicians whose names I was familiar with, in this case Martin Davis and Raymond Smullyan. By this time Raymond Smullyan was at work on what became his First-Order Logic book, and I was introduced to tableaus all over again, this time with somewhat better results. I came to appreciate the naturalness, the power, and the informativeness of tableaus.

When it came time for a thesis topic, set theory was hot. Paul Cohen had just invented forcing, Scott and Solovay came to town presenting their algebraic re-formulation, and all this heavily influenced everybody. Saul Kripke had recently introduced his possible world semantics, and in particular his semantics for intuitionistic logic. It was clear to many people that set-theoretic forcing and Kripke intuitionistic semantics were closely related. I decided to exploit that relationship by using Kripke semantics directly, carrying out the Cohen construction in an intuitionistic setting, not using countable models for set theory, but instead extracting classical independence results via the double negation embedding of classical logic into intuitionistic logic. But in order to do this properly, I felt a good foundation needed to be laid for Kripke intuitionistic semantics, for its relationship to modal semantics, for intuitionistic tableau systems, for algebraic semantics for intuitionistic logic, and so on. Thus my thesis developed in two parts: intuitionistic logic and applications to set theory. The thesis was clearly running long and my advisor, Raymond Smullyan, gave me wonderfully apt advice: "write it as a book." I did. When it was published some people read part 1, some people read part 2. I suspect few read both parts. My own future research (mostly) followed up on part 1.

By now I had developed a genuine feeling for tableau systems. I began creating or adapting them for various modal logics. What has become known as prefixed tableau systems occurred to me around this time. But I still had no particular interest in epistemic logic as such, early exposure to Hintikka notwithstanding.

At some point I came across two 1968 papers by Stalnaker and Thomason, essentially about how non-rigid designators could be modeled properly. It was a really a simplified fragment of Montague's work, though I didn't know this at the time. I quickly realized that the Stalnaker, Thomason methodology made possible Skolemization for modal logics, and eventually I even evolved a form of Herbrand's theorem. I wrote a series of technical papers exploiting all this.

Finally, I became properly aware of logics of knowledge. I met John McCarthy and heard him expound on the significance of epistemic puzzles. Fagin, Halpern, Moses, and Vardi had begun writing, separately and in various combinations, and their influence was an important one. At some point everything came together. I understood how the Stalnaker, Thomason work was the (or at least, a) good way of modeling the Fregean sense/reference distinction. I came to understand how this combined with prefixed tableaus to produce natural proof systems for logics in which de re/de dicto issues could be examined formally. I realized, much to my great surprise, that epistemic issues were a common thread running through much of what I had done over the years. So, why was I drawn to epistemic logic? In a sense, I wasn't. At a certain point I discovered that's what I had been doing all along, rather like the man who discovered he'd been speaking prose all his life.

2. What example(s) from your work, or work of others, illustrates the relevance of epistemic logic?

3. What is the proper role of epistemic logic in relation to other disciplines, for instance mainstream epistemology, game theory, computer sciences or linguistics?

I will treat these two questions as one: what is the relevance and the proper role of epistemic logic? This question is really a special case of one that is more general: what is the point of formalization. Of course there is no single, simple answer to the general question, or even to the specific one concerning epistemic logic.

Often formalization isn't what clarifies a field, it can be the other way around. If the informal foundations of a field become clear then formalization is feasible, and relatively easy. Group theory is a simple example: mathematical investigation gradually clarified the notion of a group, and once clarified, a first-order formalization becomes an exercise. Physics is an example in the opposite direction. Formalization must wait until physicists think they really know what they are talking about; right now parts don't fit together. Of course it often happens that clarification and formalization can proceed together, as was the case with set theory and the foundations of mathematics more generally.

An attempt to formalize forces us to think about what is basic and what is derivative. A good example is Hintikka's presentation of epistemic logic as modal logic. A key insight was that ignorance, not knowledge, was fundamental. An agent doesn't know something if there are possible worlds in which that something has differing truth values—multiplicity of worlds represents ignorance.

Knowledge, then, is the special case in which ignorance has vanished. This is a remarkable insight, and perhaps would not have become clear without the attempt to formalize.

Suppose we think we understand some subject, provide a formal treatment of it, create a methodology of formal proofs, and use it to derive an implausible result. What conclusions do we draw? Either our formalization is bad, and so we did not really grasp the key concepts of the area, or else our formalization is good, and there is something fundamentally problematic within the area itself. This is a negative use of formalization. It is a special case of the construction of scientific theories: formalizations are falsifiable, but only verifiable to greater and greater degrees. One bad conclusion rules out the formalization, and perhaps more with it. Finding only acceptable (and presumably useful) consequences give us confidence in the formalization, but of course nothing rules out the creation of a better formalization based on different insights. As we well know, new insights can come out of a particular formalization, and lead to its eventual abandonment. In physics, exploration of the Newtonian model helped develop insights that led to the relativistic model. In the foundations of mathematics, exploration of Frege's system helped develop the insights that led to our current set theory. It is the same with epistemic logic. The Hintikka insight gave us formal systems that are useful and applicable; exploration shows us their limitations. Perhaps the next thing is on the horizon—I'll have more to say about this below. In the meantime we can think of the Hintikka approach as akin to the Newtonian approach in physics—perhaps a bit incorrect, certainly incomplete, but useful for many things nonetheless.

Formalization, like logic in general, has two familiar sides: proof procedures and semantics. We all know this, and we all know their respective roles. Our attempts at formalization might succeed to the point where we create a proof procedure that lends itself to automation. In this fortunate circumstance we could use our proof procedure as a tool, rather like a calculator, to avoid some of the routine thinking a subject demands. This is ideal, useful for grant applications perhaps, but rare. Semantics is something different. A good, intuitive semantics psychologically convinces us we really do understand a subject, correctly or not. Modal logic provides a nice example here. Algebraic semantics was around for years, was useful for showing fundamental results about modal logics, and was capable of handling all the technical demands that were later placed on possible world semantics. Then why did possible world semantics, when introduced, change the field so profoundly? Because possible world semantics provided what seemed to be the right metaphor, the correct intuition, the satisfying feeling. One quickly found oneself talking as if modal notions really were about possible worlds, rather than as if the formalization of modal notions was about possible worlds. It's an easy distinction to blur.

Game theory, computer science, linguistics, epistemology, all are understood to some extent. There are experts in each of these fields, working away, and sometimes they are interested in foundations. Each field has its own history, methodology, intuitions. Knowledge of agents certainly plays a role in each of them, and so formal epistemology might be applied, but to what end? The answer need not be the same for the different areas just mentioned, indeed it is not even clear what "same" might mean here. I'll discuss two of the areas, computer science and game theory.

John McCarthy has famously claimed that his thermostat has knowledge. Sometimes it knows it is too cold, and sometimes it knows it is too hot. This seems fanciful, but examine the claim closely. The classic Platonic paradigm for knowledge is justified true belief. If the thermostat's beliefs about temperature weren't true, we would say it was broken and replace it, so we can assume they are true as a practical matter. The beliefs are justified—in effect the physics of the thermocouple sees to that. The essential question is, what does belief mean for a thermostat? We can give a possible world interpretation to this, but it does not seem quite natural. Rather we evaluate what a thermostat believes by what it does. It always acts on its beliefs about temperatures, and not on beliefs about anything else; this is why we have it. Acting on beliefs is more general in the world than just in the realm of thermostats. There are certainly internal beliefs that a person may have, with no visible sign of them, but generally our interest in the beliefs of a person comes from our understanding of what those beliefs will cause a person to do. This seems to be missing from the standard model that is based on possible worlds. McCarthy's example, while extreme, shows that it is plausible to talk as if automata had beliefs or knowledge. We might debate whether it is true that they do, or what it might mean if they did, but it is not unnatural to do so. Indeed, it suggests a missing element in the standard epistemic formalization: what actions does a belief entail? Is there an action-based epistemic semantics that might substitute for the possible world version?

Applications of formal epistemics come up throughout computer science. For instance, we can think of finite automata as epistemic models with transitions corresponding to agent accessibility relations. It is no coincidence that the notion of bisimulation is the same for multi-agent epistemic models and for finite automata. Thinking of automata epistemically can be fruitful, as can thinking of epistemic models as automata.

Encryption is a field crying out for proper formalization. Suppose two agents want to communicate and need to establish their identities to each other by making use of a trusted third party; in particular they may need to set up appropriate encryption / decryption keys. Protocols have been developed for such things, and correctness proofs have been given, though they have sometimes been incorrect. Formal methods have been created to try and address this problem, but the methods tend to be ad hoc and unnatural. Someday, somebody will apply formal epistemology here in a convincing and simple way.

Many varieties of formal methods have been applied to analyzing correctness of computer programs, beginning with the work of Floyd and Hoare. These generally don't look like formal epistemology, but there is a proper role for it here. In object-oriented programming, program behavior depends on the interaction of many objects. Each object has public information, perhaps also private information, and can execute various methods, that is, it can do various things when asked. Objects may or may not know about other objects, in particular which objects to ask to have certain tasks carried out, and so on. This seems like a natural area for formal epistemology. It awaits someone with the right insights.

Game theory is an area I'm currently thinking about. Knowledge plays an important role here. There are always tacit assumptions such as: each player knows what moves are possible, each player knows whose move is next, and so on. There are also epistemic assumptions that are singled out for explicit mention: common knowledge of the rationality of players is an example. Is there a full set of epistemic assumptions for some game of interest? What might it mean to be a full set? How do these assumptions function as a game proceeds? Recently I've investigated the combination of Hintikka-style epistemic logic with propositional dynamic logic; the former to represent player knowledge, the latter to represent the range of moves available. (This has been previously investigated by Van Benthem, and is related to the work of van Ditmarsch, van der Hoek, and Kooi.) There is a proof theory and a semantics, deriving from the work of Schmidt and Tishkovsky. Aumann showed that common knowledge of rationality is central to the centipede game. I can show formally, using epistemic/dynamic proofs, that common knowledge of rationality indeed determines the moves. I can also show, using semantics, that mutual knowledge of rationality does not. Does the introduction of formal epistemics into game theory add anything to Aumann's work? Perhaps. It clarifies just what is being assumed in the semi-formal work. It can ensure a claim that something, say that common knowledge of rationality is necessary, is not invalidated because some tacit condition was missed. What is personally exciting is the possibility of connecting game theory to current work on explicit versions of epistemic logic, which is discussed in questions 4 and 5.

What are my conclusions, then? Formal epistemic methods help us clarify what is basic to a field, or rather to our knowledge of and interaction with a field. In particular, which are the fundamental concepts, and which are the derivative ones. Formal epistemology may lead to proof procedures that can help us with low-level reasoning, or perhaps can be embodied in computer programs, or even in finite automata. Formal epistemics may give us a semantics that produces useful metaphors—possible worlds, for example—that help us talk to each other. But it is important that we do not conflate formal methods applied to a field with the field itself. Scientific theories of the world are good and useful, but they are not the world itself.

4. Which topics and/or contributions should have had more attention in late 20th century epistemic logic?

5. What are the most important open problems in epistemic logic and what are the prospects for progress?

The turn of a century is an artificial (though useful) marker, so questions 4 and 5 are really the same thing. I will discuss them that way. It would be interesting to see how much space other respondents devote to problems they consider important, but on which they don't intend to work. I will say up front that I intend to discuss problem areas that I think are important to the general subject of epistemic logic, are significant to me personally, and on which I do intend to work (indeed, I have already done so).

In most epistemic logics, knowledge is of *propositions*. It is not that propositions are represented directly, but rather, knowledge of a sentence generally entails knowledge of all equivalent sentences, and thus it is really propositions that are fundamental. But in mathematics, or even in everyday life, it is the *sentences* that count. If we know all equilateral triangles are equiangular, it does not follow that we know all equiangular triangles are equilateral, though these are equivalent in the Euclidean context. There are straightforward ways of getting sentences into the picture—awareness logic does this, but awareness logic machinery is really a bit of a blunt instrument. We need to get closer to sentences by more subtle means.

Justification Logics were developed near the end of the 20th century. They had their origins in a project of Sergei Artemov, to produce a constructive semantics for intuitionistic logic. The well-known Gödel-Löb logic embeds naturally in arithmetic but does not relate well to intuitionistic logic. $S4$ relates well to intuitionistic logic via any of a number of "box" translations, but it does not embed in arithmetic using the provability interpretation. The necessitation operator of Gödel-Löb logic was intended to abstract from the idea of arithmetic provability. Artemov's idea was to abstract from the idea of *explicit* arithmetic proof. (It turned out that Gödel himself had already had this idea, but it remained undeveloped and essentially unknown for a long time.) Eventually Artemov created a logic he called LP, for *logic of proofs*, in which there were infinitely many *proof terms*, with a simple family of operations on them. If t is a proof term and X is a formula, then $t : X$ is a formula, intended to be read "t is a proof of X". The logic LP does embed in arithmetic (Artemov's Arithmetic Completeness Theorem), but also $S4$ embeds in LP (Artemov's Realization Theorem). Combining all this with the embedding of intuitionistic logic in $S4$ indeed provides a constructive semantics for intuitionistic logic.

It was quickly realized that proofs were a special case of *justifications*. The Realization Theorem, connecting LP and $S4$, can be thought of as relating *explicit* and *implicit* representations of knowledge. In $S4$, taken as an epistemic logic, we might encounter a formula KX, and understand it as asserting an agent knows the proposition expressed by the sentence X. In LP we might instead encounter $t : X$, but now we must understand it as asserting that we know the sentence X itself for explicit reason t. If X and Y are equivalent sentences, and an agent knows X for explicit reason t, there will be an explicit reason, t', by which the agent knows Y, but it won't be the same reason. Generally t' will be more complex than t, embodying it and reasons for the equivalence of X and Y. (Of course there may be an independent and simpler reason for which Y is known, but this depends on circumstances.)

LP was generalized early on, so that there were explicit analogs of single agent logics of knowledge with and without factivity, positive introspection, negative introspection, and various other combinations of conditions. The resulting family became known as the family of *justification logics*. I introduced a Kripke-style semantics for them in 2005, which provides both insight into intended meanings, and admits new proofs of fundamental results such as the Realization Theorem.

Justification logics are, in a sense, the natural next thing after Hintikka-style epistemic logics. Relationships between the two families are natural, and mathematically precise. But the work is still new, and whole areas are as yet undeveloped. And now I can get to the questions I am supposed to be addressing, 4 and 5.

What counts as a justification? Better yet, what are natural operations on justifications? The only operations that have been much investigated are those that arise from formal proofs, indeed mostly from formal propositional logic proofs. These are not the only ways we justify things. There is direct observation, for instance. There is being told by a trusted source. And what else? What are natural operations on such justifications? Public (or semi-public) announcements are obvious things to consider, but work has barely begun on this. And that naturally leads me to a related issue.

So far the most successful work on justification logics has been with explicit analogs of implicit *single agent* epistemic logics. Obviously we have an interest in multi-agent versions. Some success has been achieved with multi-agent justification logics in which the justifications are understood by all agents. That is, each agent might (implicitly) know something for unspecified reasons of its own, but explicit justification terms are shared by all, and thus function as a kind of common knowledge. This works well technically, and has its applications, but it is obviously not as full a picture as we would like. It would be better if we could provide each agent with its own set of justification terms, and possibly also provide a common set. Then we might want to consider public or semi-public announcements, not just of facts but of justifications for facts. So far this remains largely undeveloped. What might be the connection between explicit announcements of justifications, and implicit announcements of facts? Probably this comes down to what form an appropriate Realization Theorem would take. What would an appropriate Kripke-style semantics be? Could we have justifications that some agents cannot understand? What about relationships between justifications and actions, as in an explicit analog of dynamic logic? This relates to my discussion of game theory above. Almost everything is open here.

I believe *explicit* epistemic logic is a major new paradigm, with a scope as broad as the whole subject of formal epistemic logic. For every development so far, for every formal epistemic research topic, we can meaningfully ask about an explicit analog. This is not just asking for something like an algebraic analog of what we do with Kripke models or for a German translation of Shakespeare. Explicit epistemic logic attempts to understand at a deeper level. It is, consequently, harder to work with, indeed much harder. The Realization Theorem has been mentioned several times now. It has both semantic and algorithmic proofs. The algorithmic proofs are very proof-theoretic in flavor, clearly having some relationship with cut elimination, but our understanding of this is still not very deep.

We have had a long, successful history of epistemic logics in which we examine what we know, what we know about what we know, how our knowledge changes with changing circumstances. I believe the work to come must address not just what we know, but why we know it, and how our reasons for knowing might change with changing circumstances. Justification logics, as we currently understand them, may be nothing more than surface exploration, but they have uncovered some promising veins to be mined.

9

Paul Gochet

Emeritus Professor

University of Liège, Belgium

---

1. Why were you initially drawn to epistemic logic?

The reading of Hintikka's epoch-making book *Knowledge and Belief* and his subsequent *Models for Modalities* aroused my interest in epistemic logic. I had been puzzled for some time by the problem raised by the formalization of hybrid inferences in which the same sentence occurs both inside and outside a verb of propositional attitude [Gochet, 1975]. Consider the following inference: "The morning star is a planet and it is known that it is a planet. Hence there is an x such that x is a planet and such that x is known to be a planet". If sentences in the scope of a verb of propositional attitude are denied their ordinary meaning and are ascribed an *oblique* meaning (Frege) or are treated as *opaque* (Quine), the two occurrences of "x" following "such that" become *equivocal*. Hence they cannot be quantified over by the same quantifier "there is an x". To this intriguing problem, Hintikka offered a solution which struck me as the best one. For him, the variable "x" takes ordinary individuals as values in the two occurrences. But the inference is enthymematic. It is valid with the proviso that it is strengthened by appropriate auxiliary premise. On Hintikka's construal the inference can be formalized in this way:

Premise: $P(a) \wedge K_b(P(a))$
Auxiliary premise $\exists x:(x = a \wedge K_b(x = a))$
Conclusion: $\exists x : P(x) \wedge K_b(P(x))$

The problem of hybrid contexts will reappear in another guise in connection with the formalisation of *knowing how*. See Section 4.

In the late eighties Professor Pierre Wolper and I launched a research project on epistemic logic supported by the Ministry of

Education. We hired a computer scientist (Philippe Lejoly) and a philosopher (Eric Gillet) to work on the subject. Let we briefly review some of the findings of these two researchers.

In many applications of the logic of knowledge, the notion of knowledge is an *external one*. Lejoly rather takes the point of view of the agent, i.e. he advocates an *internal* or *subjective* approach. He provides each agent with a model of his or her own knowledge which he or she uses to evaluate formulas and which he or she updates when receiving new information. Lejoly's innovation consisted in substituting finite knowledge structures to the infinite structures used by Halpern and Moses. This required of him to introduce circularity or self-reference into the definition of a model.

An agent's model, on Lejoly's account, consists of two parts, knowledge of the environment and knowledge about the knowledge of other agents. The satisfaction relation is relativized to each agent. The satisfiability of $\neg\varphi$ does not boil down to the non-satisfiability of φ. The formula φ may be neither satisfied nor not satisfied. This happens, Lejoly observes, when the agent is *not aware of* φ, a situation which is captured by truth-value gaps, i.e. partiality, in semantics. The partiality which characterizes the agent's models is absent from the *global model*. The latter, which is a conceptual tool inaccessible to a particular agent, consists of a complete truth-assignment representing the "real" environment together with the individual models of the agent.

An interesting result can be obtained in this setting : negative introspection holds at the level of the individual agent's model but it does not hold at the global level. The agent, who is *ex hypothesi* capable of checking the truth of a formula in his or her own individual model, is entitled to say, if he comes to a negative conclusion, "If I, agent i, do not know φ, then I know that I do not know φ". Now consider an agent who is not aware of the primitive proposition p. In this case both $\neg K_i p$ and $\neg K_i \neg p$ are satisfied at the global level, but $\neg K_i p$ is not satisfied on agent i's model. So $K_i \neg K_i p$ is not satisfied at the global level , although $\neg K_i p$ is. Hence negative introspection fails at the global level [Lejoly 1991].

The problem of logical omniscience which bedevils epistemic logic has been given a detailed treatment by Ronald Fagin and Joseph Halpern in "Belief, Awareness and Limited Reasoning" [Fagin and Halpern 1988]. Eric Gillet , who started his research by a thorough study of that paper, suggested to replace the *awareness* predicate by a *depth analysis* function symbol [Gillet 1992, Gochet

1992, Gillet and Gochet 1993, Gochet and Gillet 1996].

It is well known that a piece of reasoning expressed in natural language can receive coarse-grained or fined-grained analyses. What degree of depth analysis should we prefer ? Coarse analysis may be insufficient to guarantee the validity of the inference. But once enough formal structure has been exhibited to secure validity there is not point in looking for an even more fine grained formalization. The optimal formal representation is the one which reveals the least structure which is enough to secure validity (Quine's *maxim of shallow analysis*) [Haack 1978].

Consider the normality axiom $K_i\varphi$, $K_i(\varphi \supset \psi) \Rightarrow K_i\psi$. Any attempt to restrict this axiom in order to block the unwanted consequence of logical omniscience commits us either to restrict the agent's logical rationality or to impose arbitrary restrictions. Depth analysis avoids the dilemma. The agent is perfectly rational. If he or she does not infer some conclusion deducible from the premises, the trouble lies in his or her failure to go deep enough in the analysis of the logical structure of the formula.

Gillet defines the notion of *depth analysis* A as a mapping which associates a formula $A^i(\varphi)$ with a pair formed by a formula φ and an integer i in accordance with four requirements. I will only mention here the last two requirements:

1. subsumption: if $i \leq j$ then $A^i(\varphi)$ subsumes $A^j(\varphi)$

2. monotonicity: if $i \leq j$ then $A^i\ A^j(\varphi) = A^i(\varphi)$

[Gillet proved that the result of any depth analysis of an initial sentence is independent of the number and order of steps processed before reaching the level of depth needed to insure the validity of the inference.]

Let us restrict ourselves to an example belonging to propositional logic. Roughly speaking, the *depth of a formula* ($D\varphi$) is the *maximal number* of nested connectives in the syntactic tree of the formula. Take for instance the epistemic axiom mentioned above in which the depth of the constituents is indicated by numbers allocated to the connectives occurring in them:

$$\varphi: \quad p \underset{1}{\supset} q$$

$$\varphi \supset \psi: \quad (p \underset{1}{\supset} q) \underset{4}{\supset} (p \underset{3}{\supset} ((q \underset{1}{\supset} r) \underset{2}{\wedge} (r \underset{1}{\supset} q)))$$

$$\psi: \quad p \quad \supset \quad ((q \quad \supset \quad r) \quad \wedge \quad (r \quad \supset \quad q))$$
$$ 3 1 2 1$$

The depth of $\varphi = 1$, the depth of $\varphi \supset \psi = 4$ and that of $\psi = 4$. If the agent analyzes the reasoning at the depth level 4, he or she is safe. The inference is valid. This is no longer the case if he or she content him or herself with a lower depth of analysis, let say level 3. In that case, the underlined constituents of formula 2 are treated as atoms and formula 2 acquires the form, $(p \supset q) \supset (p \supset (s \wedge t))$. It is obvious that $p \supset q$ and $(p \supset q) \supset (p \supset (s \wedge t))$ together with $p \supset (q \supset r) \wedge (r \supset q))$ cease to instantiate *modus ponens* .

Within *possible worlds semantics* it is easy to construct a model which verifies $K(p \supset q)$ and $K((p \supset q) \supset (p \supset (s \wedge t)))$, but falsifies $K(p \supset ((q \supset r) \wedge (r \supset q)))$. We have to take up the following truth value assignments at world $\boldsymbol{w}$: $\Im(p, \boldsymbol{w}) = $ T, $\Im(q, \boldsymbol{w}) = $ T, $\Im(r, \boldsymbol{w}) = $ F, $\Im(s, \boldsymbol{w}) = $ T, $\Im(t, \boldsymbol{w}) = $ T [Gillet 1992],[Gillet & Gochet 1993], [Gochet 1995], [Gochet & Gillet 1996].

Around 2006, Professor Pascal Gribomont and I worked out a survey of epistemic logic covering the research done on the subject in the second half of the 20th Century. Multimodal and multi-agent systems were presented together with recent developments such as doxastic dynamic and epistemic dynamic logic. The paper ends with a section devoted to applications to software engineering . Subsection 8.1 deals with *Logic in Concurrent Programs*, Subsection 2 with *Knowledge in Asynchronous Message Passing Systems*, Subsection 8.3 with *Some Applications of Common Knowledge* and Subsection 8.4. with *Knowledge bases* [Gochet & Gribomont 2006].

Later on I became interested in the formalization of *knowing how*. I compared the possibilities and limits of various formalisms set up, mainly in the A.I.community, to deal with the problem. and devoted an essay to the topic [Gochet 2007]. In that essay I compare three approaches: (1) the formalization of knowing how in Robert Moore's *A Formal Theory of Knowledge and Action* [Moore 1985,1995], (2) the formalization of knowing how within the framework of multimodal propositional logic designed to study the logical relations between knowledge, ability, result and opportunity (the KARO program) [van Linder, van der Hoek, Meyer 1998], [Meyer, de Boer, van Eijk, Hindriks, van der Hoek, 2001], (3) the formalization of knowing how in the Situation Calculus [McCarthy and Hayes 1969, 1998] More recently, I widened my field of inquiry and started to explore the logic of intentionality [Gochet 2010].

2. What example(s) from your work, or work of others, illustrates the relevance of epistemic logic?

Initially the logic of "see to it that" (*stit*-logic) developed by Belnap, Horty and others was meant to elucidate the logic of human action. In that context, acting is conceived as interfering with a course of events by one's choice. *Stit logic* gave rise to an abundant literature which is of importance not only for logic but also for metaphysics. As Belnap and his co-authors observe the structure used to interpret *stit*-formulas are not just mathematical curiosities but "they describe—up to legitimate idealization—the world in which agents act. [Belnap, Perloff and Xu 2001]

Recently it has been shown by several authors working independently (Herzig and Troquard, Semmling and Wansing , Broersen, Lorini and Schwarzentruber) that the expressive power of the *stit*-operator could be significantly increased if it were combined with the *epistemic operator* K. *Stit*-logic and epistemic logic can be frutifully combined.

In moral philosophy it is highly important to distinguish *intentionally doing* φ from *knowingly doing* φ. The first entails the second but not conversely. The second is closed under causal implication but the first is not. If we want to state these properties by axioms in a formal system, our fist task is to provide a formal rendering of these adverbial constructions. For that purpose, Jan Broersen created a new operator made out of the K-operator and the *stit* operator glued together.

In the well-known BDI architecture [Rao and Georgeff 1991], *belief, desire* and *intention* are taken as primitives and axioms which relate them are stated. Emiliano Lorini and François Schwarzentruber go one step further [Lorini and Schwarzentruber, forthcoming]. They define *desirability of a state of affairs* φ *by an agent* i in terms of the knowledge that φ is good for agent i. Here again the K-operator and the evaluative operator "It is good that" are glued together in the definition: $DES_i\varphi =_{def} K_iGOOD_i\varphi$. At a later stage, the K-operator is used to formalize the intentional verb *regret*.

Lorini and Schwarzentruber spell out the following definition in which CHP_i stands for "agent i could have prevented φ": $REGRET_i\varphi =_{def} K_iCHP_i\varphi$. With these definitions at their disposal, they can ensure the validity of an implication which links together *regret, knowledge* and *desire*: $REGRET_i\varphi \supset (K_i\varphi \wedge DES_i\neg\varphi)$.

Two other authors, Caroline Semmling and Heinrich Wansing,

who succeeded in combining the BDI architecture with the *stit* logic, have observed an unnoticed asymmetry between *belief* and *desire* on the one hand and *knowledge* and *intention* on the other. Beliefs can conflict with one another . Desires can conflict too but intentions and pieces of knowledge cannot [Semmling and Wansing 2008]. The authors' example brings that out quite clearly. I may both *desire* to donate one of my kidneys to save my brother, and *not desire* to donate it as it will affect my physical integrity. Yet I do not entertain the incoherent desire of *donating* and *not donating* one kidney.

To capture this situation we need a modal logic in which $(\Box p \wedge \Box q) \supset \Box(p \wedge q)$ does not hold. Such a logic exists, but it requires of us that we switch from Kripke's *relational semantics* for modal logic to Montague Scott's *neighbourhood semantics* [Chellas1980,210, Straetmans 1991]. Semmling and Wansing take advantage of this possibility. The semantics appropriate for beliefs and desires is the second one. The semantics appropriate for knowledge and intentions is the first one.

Here I would like to stress that the illuminating observation made by Semmling and Wansing can be extended to *deontic logic*. Besides *desires* and *beliefs*, *duties* can conflict. .Here is an example due to the casuist Raymond of Penaford quoted by Frédéric Sart: "The casuist imagines a householder who is asked by people bent on taking the life of someone hiding in the house whether she is in. Replying "Yes" is obligatory by virtue of the duty to tell the truth, but at the same time it is forbidden by virtue of the duty to save a life [Sart, 2009, 132]".

Suppose we adopt the law $(\Box p \wedge \Box q) \supset \Box(p \wedge q)$ for the deontic operator Op [it is obligatory that p]. We can reason as follow: If in a given situation it is obligatory to tell the truth and forbidden to tell the truth (i.e. obligatory to lie), then it is obligatory to tell the truth and to lie. But telling the truth and lying is impossible. As doing what is impossible is not obligatory (in virtue of the principle *ought implies can*), it follows, by *modus tollens*, that it is not obligatory to tell the truth and to lie.

This conclusion is ethically unbearable. It is rejected by Nicolai Hartmann in his *Ethik*. He holds that when we are confronted to a conflict of duties, we are guilty whatever we do. As he puts it "The alternative is here not between wrong and right, but between wrong and wrong. Either way a value is violated and, again, either way a value is fulfilled. Whoever stands in such a predicament—and life continually places man in such—cannot escape without offence [Hartmann 1926, transl. 1932, 1950, 301].

To block the chain of inferences which leads to the counterintuitive ethical conclusion denounced by Hartmann, we need again to adopt a semantics which deprives the modal formula quoted above from its validity. This can be done for free by using again the neighbourhood semantics that was adopted to cope with the logic of desire.

What emerges out of this is that *knowledge, belief, desire, regret, obligation, permission, past* and *future* belong to a complex *network of modalities* in which knowledge occupies a pivotal position, as shown by its being an essential ingredient of many other modalities such as *desire, regret* and *knowingly doing*.

3. What is the proper role of epistemic logic in relation to other disciplines, for instance mainstream epistemology, game theory, computer sciences or linguistics?

In the first part of this section a plea will be made for the claim that epistemic logic can be a very useful tool to tackle philosophical problems which belong to mainstream epistemology. The example I have chosen is the principle discussed by Timothy Williamson in *Knowledge and Its Limits* under the name "principle of weak verificationism". The principle reads as follows "every truth is knowable", formally $\forall p \ (p \supset \Diamond Kp)$ [Williamson 2000]. For our purpose the quantifier can be dropped and the principle can be formally rendered by $p \supset \Diamond Kp$.

In a paper published in 1963, applying uncontroversial laws of modal logic to the verification principle, Fitch derived a paradoxical conclusion "every truth is known", formally $p \supset Kp$. Some authors rightly call it the *collapse principle* since, combined with the T axiom, it leads to $p \equiv Kp$ where K has been turned into an empty symbol.

Many proposals have been made to circumvent the so-called Fitch paradox. [For a characterization of those proposals see Johan van Benthem's "What one may come to know" [van Benthem 2004]]. I will focus on the solution offered by Alenxander Costa Leite. Relying upon the methodology developed for combining logics by Dov Gabbay and his co-authors [D.Gabbay, A.Kurucz, F.Wolter and M. Zakharyaschev 2003], Kosta-Leite argues that the correct minimal logic we need to state the paradox is composed by a *fusion* of modal languages and logics and by a *fusion* of modal frames. Having done that, he sets up a counter-model to Fitch's paradox in which $p \supset \Diamond Kp$ is true, but $p \supset Kp$ is false.[Costa-Leite 2006, 2007].

This semantic refutation of the Fitch paradox is intriguing, It suggests that there must be a flaw somewhere in the derivation which leads to the formula $(p \supset \Diamond Kp) \supset (p \supset Kp)$. Where is the flaw? To answer this question we have to stick to the logic used by Costa Leite, namely a bimodal logic which contains the alethic modal operator $\Box$, the epistemic operator K, the modal axiom licensing the distribution of $\Box$ over $\supset$ and, finally, the standard axioms of epistemic logic for K. The rule of necessitation is extended. It also applies to the theorems of epistemic logic.

With this apparatus, we can reconstruct the derivation of the collapse principle form the weak verification principle in this way [Gochet 2008]:

1. $(p \wedge \neg Kp) \supset \Diamond[K(p \wedge \neg Kp)]$ instance of the weak verification principle

2. $K(p \wedge \neg Kp) \supset \bot$ theorem of epistemic logic (See Gochet & Gribomont 2006)

3. $\neg K(p \wedge \neg Kp)$ 2, *Modus tollens*

4. $\Box \neg K(p \wedge \neg Kp)$ 3. Necessitation

5. $\neg \Diamond [K(p \wedge \neg Kp)]$ 4. Definition

6. $\neg (p \wedge \neg Kp)]$ 1,5, *Modus tollens*

7. $p \supset \neg Kp$ 6. Definition

The flaw occurs in (1). As the verification principle is not a theorem of logic, the substitution which leads from its canonical form to the instance (1) is not logically warranted. We are not better off if we take (1) as an auxiliary hypothesis in a conditional proof. Substitution on hypotheses is not allowed.

Of course, there are other derivations, much more refined and sophisticated [Balbiani, Baltag, van Ditmarsch, Herzig, de Lima, Hoshi 2007] which are immune to this criticism but they are irrelevant to the problem at hand. I restricted myself to the proof theory that fits Costa-Leite's model theory for bimodal logic. My aim here was to show how epistemic logic can be part and parcel of a philosophical argument dealing with a typical problem of mainstream epistemology: the limits of knowledge.

Another example of the relevance of epistemic logic to epistemology can be found in Vincent F. Hendricks' book *Mainstream and Formal Epistemology* [Hendricks 2006]. In that book, however the other side of the coin, i.e. the relevance of epistemology to logic, is also scrutinized. The book has been written, as Peter Østrøm puts it, in the grand old tradition of trying to combat skepticism and to assure that human beings can in fact possess knowledge. The problem is old but the approach is new. The author developed a formal language and an axiomatic system powerful enough to express new epistemological concepts.

Looking at the standard axioms of epistemic logic, Hendricks raises this fundamental question "Which epistemic axioms may be validated by an operator based on the definition of limiting convergent knowledge for discovery methods?". The agent indices associated with the K-operator in Hendricks' formal language are viewed as arguments of discovery functions. This changes the significance of the T axiom of epistemic logic "$Kp \supset p$". It no longer ascribes the passive property of truth to known propositions but the active property of resistance to objections, i.e. *indefeasibility*.

A striking example of the relevance of epistemic logic to epistemology can be found in the eighth chapter of the book. In that chapter the author spells out a new *model of inquiry* obtained by mixing alethic, tense and epistemic logics with a few motivational concepts drawn from computational epistemology.

The concept of knowledge underlying this model is that of limiting convergence. Important epistemological distinctions such as the distinction between *temporal necessity*, *universal necessity* and *empirical necessity* are recast as logical operators for which a formal semantics (truth-conditions) is provided. A crucial question is raised concerning the validity of the axioms of epistemic logic: does this validity depend on enforcing methodological recommendations or does it depend on structural features of the learning mechanism?

A useful distinction is drawn by the author between synchronic and diachronic axioms. An epistemic or doxastic or combined implicational axiom is synchronic if the consequent obtains at the very same time the antecedent obtains. It is diachronic if the consequent obtains later or would have obtained later than the antecedent if things had been otherwise. In the light of this distinction, a nuanced answer can be given to the question whether the axioms K,T, 4 and 5 are valid or not.

In modal operator epistemology, axiom K and T are valid *synchronically*, axiom 4 is valid *diachronically* and axiom 5 is not valid either way. These verdicts of validity or lack of validity receive a detailed justification. For instance the author argues that axiom 4 ($K_\delta h \supset K_\delta K_\delta h$) holds if the method δ has the property that he calls "consistent expectations". Among *unexpected epistemological results* obtained by the author, let me mention just this one: a discovery method cannot entertain perfect memory and consistent expectation at the same time.

4. Which topics and/or contributions should have had more attention in late 20th century epistemic logic?

Over the last twenty years of the 20th Century major contributions have been made to the study of *knowing how* both by A.I. logicians and by philosophers, but, due to a lack of interaction between these two communities, some problems have not been addressed. What has been missing is a *logico-philosophical* account of the epistemic construction " knowing how ".

Robert Moore observes that an agent can know how to achieve a state of affairs p by performing a complex action A without knowing from the very beginning what he is to do at every step. "All that is necessary is " that the agent know what to do first and that he will know what to do at each subsequent step [Moore 1985, 1995]. An additional requirement was pointed out by Munindar Singh: it is important to require that when the agent finally achieves the condition which is his or her goal, he or she knows that the goal has been reached, because this helps limit the actions selected by the agent. As Singh puts it "If p holds, but the agent does not know this, then he might select still more actions to achieve p [Singh 1994]".[In his studies on the ethology of birds, Conrad Lorenz has shown the survival value of knowing that the goal pursued has been achieved. We owe this to Professor Crubbelier, University of Lille 3].

In a later paper, Singh offers the following equivalence as a characterization of *knowing how*. Agent x *knows how* to bring about the state of affairs (or the event) p if and only if either agent x knows that p is already the case or there is an action a such that at some scenario x knows that a is being performed and that for all scenarii if a is performed, x knows how to bring about p.This characterization of knowing how is encapsulated in the formula below [Singh 1999].

$$K_t p \vee (\bigvee a : K_t(\mathbf{E} < a > true \wedge \mathbf{A}[a]K_h p) \Leftrightarrow K_h p \text{ [Singh 1999]}$$

One can know how to write a novel *(accomplishment)* or how

to walk (*action*). Let us focus on an intermediate case : knowing how to perform a certain dance on the stage. This kind of knowing how belongs to the sort of knowing how which Frank Lihoreau calls *ability-entailing knowing-how*.

A problem arises here if we combine Singh's logical account of knowing how with Lihoreau's conceptual analyses. *Knowing that* creates an *intensional context*. Coreferential expressions cannot be substituted *salva veritate* within the scope of knowing that. On the contrary, abilities and ability-entailing knowing-how are *extensional*.

The following example due to Carr and discussed by Lihoreau brings that out clearly. If a famous dancer was to perform before an audience an item of his repertoire which he has himself given the title 'A performance of Improvisation No.15' and if the movement of the dance turned out to be, unknown to the dancer, a perfect replica of the movements of the semaphone version of Gray's 'Elegy', we would be ready to say that , in the ability-entailing sense of knowing how, the dancer *knows how* to perform a semaphone version of Gray's 'Elegy'. But we would not be ready to say that the dancer *knows that* he is performing a semaphone version of Gray's elegy.

The contrast between the intensional nature of the operator *knowing that* and the extensional nature of the operator *knowing how* [in the ability-entailing sense] raises a serious problem for Singh's characterization of knowing how since it allows the same variable "p" to occur both in the scope of K_t and in the scope of K_h. Equivocation is in the offing. Equivocation can be avoided if we read "p" as a schematic letter or place holder which cannot be quantified over.

5. What are the most important open problems in epistemic logic and what are the prospects for progress?

For a long time epistemic logic has been a subdomain of modal logic. An alternative view came to the foreground when Richard Scherl and Hector Levesque supplemented the Situational Calculus, a dialect of first-order logic, by adding an *epistemic component* to it which allows the formalization of constructions such as "agent x knows that φ as a result of performing informative action a". The question arises whether one of the two competing formalisms (epistemic logic and situation calculus) will eventually supersede the other. This is, in my opinion , an *important open problem*.

The situation calculus is a *multi-sorted* first-order logic designed to represent and reason about changes. There are two distinguished *sorts* of first-order terms: *actions* and *situations*. Situations are possible world histories or, more precisely, results of successive actions. Both actions and situations can be quantified over. Besides standard predicates and function symbols, the situation calculus contains predicates and function symbols with a situation variable or constant appended to them.. These denote *fluents*, i.e. relations and functions whose values may vary from situation to situation.

Our confidence in our plans depends on the assumption that most of our actions are independent so that they leave unaffected a large portion of the world.. To borrow an example from Richmond Thomason [Thomason 2009], we would cease to use a word processor if whenever we type a letter we automatically erased another. The situation calculus captures this kind of *inertia* by frame axioms, i.e. by axioms which limit or "frame" the effects of actions. It behoves the proponents of the Situation calculus to find a method which keeps the number of frame axioms needed under control. This problem, known as the *frame problem*, has received a concise and perspicuous solution [Brachman and Levesque 2004]..

In epistemic logic, the epistemic operator $K\varphi$ is *characterized* by axioms at the level of the *object language* (K, T, 4, 5). It is *defined* at the level of *metalanguage*. Possible worlds and accessibility relations appear only in the metalanguage. On the contrary, in the situation calculus, situations and accessibility relations occur at the level of the object-language. They are said to be "reified". For instance, the binary predicate $K(s''$, $do(sense_F, s)$ can be rendered by "a situation s'' is accessible after action *sensing whether F* has been performed in situation s". The predicate just quoted denotes the very accessibility relation which serves to define knowledge in epistemic logic.

In 1993 Richard Scherl and Hector Levesque spelled out an axiom designed to capture the meaning of that predicate [Scherl and Levesque 1993]. The axiom reads as follows :

$$K(s'', do(sense_F, s)) \equiv \exists s'.\ s''\ do(sense_F, s') \\ \wedge K(s', s) \wedge [F(a, s') \equiv F(a, s)]$$

In natural language this means that "the situations K-related to $do(sense_F, s)$ are precisely the images, under the action *sense* $_F$, of those situations K-related to s that assign the same truth value to F as does s. [Reiter 2001]". As Levesque and Lakemeyer put it, the effect of this axiom is, roughly, that it eliminates from further consideration all those situations which disagree with the result of *sensing*.[Levesque and Lakemeyer 2008].

Commenting on their axiom in an expanded version of their essay, Scherl and Levesque wrote that their axiom had no formal representation in (propositional) epistemic logic.[Scherl and Levesque 2003]. This statement amounts to claiming that the situation calculus is an autonomous and indispensable formalism. Recent developments however invite us to mitigate that claim.

In 2004 Lakemeyer and Levesque revised the situation calculus by eliminating situations and possible worlds from the object-language of the calculus. They replaced quantification over all situations by the introduction of the □ modal operator. This looked like a step toward modal logic. But it was not.

The conciseness and perspicuity of the solution to the frame problem alluded to above partly depends on *the ability to quantify over actions*. This ability is preserved in the revised situation calculus worked out by Lakemeyer and Levesque. The two authors cease to quantify over situations but continue to quantify over actions.

The multimodal propositional logic usually presented as an alternative to the situation calculus is a combination of dynamic logic and epistemic logic. In dynamic logic, actions are treated as indices but cannot be quantified over. Hence the kind of multimodal logic proposed as a substitute for the situation calculus cannot measure up to the situation calculus with respect to the frame problem, Scherl and Levesque seem to win.

Recently however the positions have changed. In "Optimal Regression for Reasoning about Knowledge and Actions", Hans van Ditmarsch, Andreas Herzig and Tiago de Lima have shown how, in the propositional case, both Reiter's and Scherl & Levesque's solutions to the frame problem can be modelled in *dynamic epistemic logic* [van Ditmarsch, Herzig, de Lima 2007].

Dynamic epistemic logic (DEL) should not be mistaken for the combination of dynamic logic with epistemic logic mentioned above. DEL is an entirely new logic which, among other things deals with public announcements [see van Ditmarsch, van der Hoek, Kooi, 2008].

A successful reconciliation of the situation calculus with epistemic logic had already been carried out by Robert Demolombe and Pilar Pozos Parra, with due consideratiion for the complexity questions [Demolombe and Pozos Parra 2000]. The three authors however go further. They insist on providing *optimal decision procedures* for their formalism.

These recent developments lend support to the *Convergence Thesis* advocated by Johan van Benthem in "Modal Logic Meet Situation Calculus". The two rival formalisms will share more and more features.

Acknowledgements

In writing this paper I received much help from discussing on the spot with the Research Team in Logic directed by Prof. Shahid Rahman in the University of Lille 3 and later with the Research Team in Logic directed by Professor Angel Nepomuceno and Senior Researcher Hans van Ditmarsch in the University of Sevilla. I thank very much the members of those teams.

References

Belnap, Nuel, Perloff Michael, Xu Ming 2001, *Facing the Future. Agents and Choices in Our Indeterminist World*. Oxford, OUP.

van Benthem, Johan. 2004, "What one may come to know", *Analysis*, 64, 95-105.

van Benthem, Johan 2007, "Situation Calculus meets modal Logic". Technical Report PP-2007-02 ILLC, Amsterdam.

Brachman, Ronald and Levesque Hector 2004, *Knowledge, Representation and Reasoning*, Amsterdam, Elsevier.

Broersen, Jan forthcoming "An *xstit*-Logic Analysis of Intentional Action".

Chellas, Brian 1980, *Modal Logic, an Introduction*, Cambridge, CUP.

Costa, Leite Alexander 2006, "Fusions of Modal Logics and Fitch's Paradox", *Croatian Journal of Philosophy*, 6, 281-290.

Costa, Leite Alexander 2007, *Interactions of Metaphysical and Epistemic Concepts,* Neuchâtel, Neuchâtel University PhD.

Demolombe, R. and M.P.Pozos-Parra 2000, "A simple and tractable extension of situation calculus to epistemic logic". In Z.W.Ras and S.Ohauga, eds. Proceedings of the 12th *International Symposium ISMIS, LNAI* 1932, Berlin, Springer.

van Ditmarsch, Hans, Herzig Andreas and de Lima Tiago 2007, "Optimal Regression for Reasoning about Knowledge and Action" *Association for the Advancement of Artificial Intelligence*, 1070-1075.

van Ditmarsch, Hans, van der Hoek Wiebe, Kooi Barteld 2008, *Dynamic Epistemic Logic*, Dordrecht, Springer.

Fagin, Ronald and Halpern Joseph, "Belief, Awareness and Limited Reasoning" *Artificial Intelligence*, 34, 39-76.

Gabbay, O., Kurucz A., Wolter F., Zakharyaschev M. 2003, *Many-Dimensional Modal Logics. Theory and Applications.* Amsterdam Elsevier.

Gillet, Eric, 1992, *Contributions à la logique de la connaissance. Omniscience logique et connaissance faible.* Liège, Université de Liège PhD.

Gillet, Eric and Gochet Paul 1993, "Le problème de l'omniscience logique", *Dialectica*, 143-171.

Gochet, Paul 1972,1975 "La logique du sens" in *Le Langage, le Sens et l'Histoire*, Lille, Publications de l'Université de Lille 3, 225-240.

Gochet, Paul 1995, "Problèmes de logique de la connaissance" In *Méthodes logiques pour les sciences cognitives* edited by Jacques Dubucs and François Lepage, Paris, Hermès, 281-294.

Gochet, Paul and Gillet Eric 1996, "A New Approach to Logical Omniscience" in *Calculemos ... Matematicas y Libertad*, ed. by J.Echevarra, J. de Lorenzo, Lorenzo Pena, Madrid Editorial Trotta, 321-334.

Gochet, Paul, "Un problème ouvert en épistémologie: la formalisation du savoir faire" in *Logique épistémique et Philosophie des Mathématiques* ed. by Thierry Martin and Philippe Mongin, Paris, Vuibert, 3-37.(with comments by Michael Cozic, Paul Egré and Gabriel Sandu).

Gochet, Paul and Gribomont Pascal, 2006, "Epistemic Logic", in Dov Gabbay and John Woods Eds , *Handbook of the History of Logic*, vol.7, 99-195.

Gochet, Paul, 1998, "Note sur le paradoxe du connaissant" in *Linguista sum*, ed. by E.Danblon, M.Kissine, F.Martin, Chr.Michaux and S.Vogeleer, Paris, L'Harmattan.

Gochet, Paul, 2010, "The Logic of Intentional Constructions" forthcoming in *Balkan Journal of Philosophy*.

Haack, Susan 1978, *Philosophy of Logics*, Cambridge, CUP.

Hartmann, N. 1926, 1950, *Ethik*, *Ethics*, vol.1, transl. by Stanton Coit, London, Allen and Unwin.

Herzig, Andreas and Troquard Nicolas 2006, "Knowing How to Play. Uniform Choices in Logics of Agency" in *Agents and Multi-Agent Systems* (AAMAS) ed. by G.Weiss and P.Stone, Hakodate Japan ACM, 209-216.

Hendricks, Vincent F. 2006, *Mainstream and Formal Epistemology*, Cambridge, CUP.

Hintikka, Jaakko 1962, 2005, *Knowledge and Belief. An Introduction to the Logic of the Two Notions*, prepared by Vincent F. Hendricks and John Symons, Cornell University.

Hintikka, Jaakko 1969, *Models for Modalities*, Dordrecht, D.Reidel Publishing Co.

Lejoly, Philippe 1991, "A Subjective Logic of Knowledge", *Logique et Analyse*, International Symposium on Epistemic Logic, 121-132.

Levesque, Hector and Lakemeyer Gerhard 2008, "Cognitive Robotics" *Handbook of Knowledge Representation*. Amsterdam, Elsevier.

Lihoreau, Frank 2008, "Knowledge-How and Ability" in *Knowledge and Questions* edited by Frank Lihoreau, *Grazer Philosophische Studien*, 2008, 77, 263-305.

van Linder, B. van der Hoek W., Meyer J.J.Ch. 1998, "Formalizing Abilities and Opportunities of Agents" *Fundamenta Informaticae*, 34, 53-101.

Lorini, Emiliano and Schwarzentruber François, "A Logic for Reasoning about Counterfactual Emotions" *forthcoming*.

McCarthy, J. and Hayes P.J. 1969, "Some Philosophical Problems from the Standpoint of Artificial Intelligence", in McCarthy John 1988, *Formalizing Common Sense* ed. by Vladimir Lifschitz, Exeter, Intellect, 21-63.

Meyer, J.J.Ch., de Boer F., van Eijk R. Hendriks K. and van der Hoek W. 2001, "On Programming Karo Agents" *Logical Journal of the IGPL*, 9, 245-256.

Moore, Robert 1985,1995 , "A Formal Theory of Knowledge and Action" in *Logic and Representation*, Stanford, CSLI Publications 27-70.

Rao, Anand and Georgeff Michael 1991, "Modeling Rational Agents within a BDI-Architecture ".in *Principles of Knowledge Representation and Reasoning* edited by James Allen, Richard Fikes, Erik Sandewall, San Mateo, Morgan Kaufmann, 473-484.

Reiter, Raymond 2001, *Knowledge in Action : Logical Foundations for Specifying and Implementing Dynamical Systems*, Cambridge Mass., The MIT Press.

Sart, Frédéric 2009, "A Purely Combinatorial Approach to Deontic Logic", *Logique et Analyse*, 206, 131-138.

Scherl, Richard and Levesque Hector 1993, "The Frame Problem and Knowledge Producing Actions" in *Proceedings of National Conference of Artificial Intelligence*, AAAI Press.

Scherl, Richard and Levesque Hector 2003, "Knowledge, Action and the Frame Problem" *Artificial Intelligence* 144.

Semmling, Caroline and Wansing Heinrich 2008, "FROM *bdi* AND *stit* TO *bdi-stit* LOGIC", *Logic and Logical Philosophy*, 17, 180-211.

Singh, Munindar P. 1994, *Multiagent Systems. A Theoretical Framework for intentions, Know-How and Communication*, LNAI 799, Berlin, Springer Verlag.

Singh, Munindar P. 1999, "Know-How" in *Foundations of Rational Agency* edited by Michael Wooldridge and Anand Rao, Dordrecht, Kluwer Academic Publishers, 105-132.

Thomason, Richmond 2009, "Logic and Artificial Intelligence", In *The Development of Modern Logic*, edited by Leila Haaparanta, Oxford, OUP, 848-902.

Williamson, Timothy 2000, *Knowledge and its Limits*, Oxford OUP.

10
Joseph Y. Halpern

Professor
Department of Computer Science
Cornell University, USA

Why were you initially drawn to epistemology (and what keeps you interested)?[1]

I was initially drawn to epistemology because I enjoyed looking at puzzles that involved epistemic reasoning. Perhaps my favorite is the *muddy children puzzle*, which is a variant of the well known "wise men" or "cheating wives" puzzles [Gamow and Stern 1985; Gardner 1984; Littlewood 1953]. The following version is taken almost verbatim from [Barwise 1981].

> Imagine n children playing together. The mother of these children has told them that if they get dirty there will be severe consequences. So, of course, each child wants to keep clean, but each would love to see the others get dirty. Now it happens during their play that some of the children, say k of them, get mud on their foreheads. Each can see the mud on others but not on his own forehead. So, of course, no one says a thing. Along comes the father, who says, "At least one of you has mud on your forehead," thus expressing a fact known to each of them before he spoke (if $k > 1$). The father then asks the following question, over and over: "Does any of you know whether you have mud on your own forehead?" Assuming that all the children are perceptive, intelligent, truthful, and that they answer simultaneously, what will happen?

---

[1] Reprinted from *Epistemology: 5 Questions*, edited by Vincent F. Hendricks and Duncan Pritchard. New York, London: Automatic Press / VIP, 2008: 155–166.

There is a "proof" that the first $k-1$ times he asks the question, they will all say "No," but then the kth time the children with muddy foreheads will all answer "Yes."

The "proof" is by induction on k. For $k=1$ the result is obvious: the one child with a muddy forehead sees that no one else is muddy. Since he knows that there is at least one child with a muddy forehead, he concludes that he must be the one. Now suppose $k=2$. So there are just two muddy children, a and b. Each answers "No" the first time, because of the mud on the other. But, when b says "No," a realizes that he must be muddy, for otherwise b would have known the mud was on his forehead and answered "Yes" the first time. Thus a answers "Yes" the second time. But b goes through the same reasoning. Now suppose $k=3$; so there are three muddy children, a, b, c. Child a argues as follows. Assume that I do not have mud on my forehead. Then, by the $k=2$ case, both b and c will answer "Yes" the second time. When they do not, he realizes that the assumption was false, that he is muddy, and so will answer "Yes" on the third question. Similarly for b and c.

The argument in the general case proceeds along identical lines.

Let us denote the fact "at least one child has a muddy forehead" by p. Suppose that in fact there are 5 muddy children. Then even before the father speaks, each child knows p. So it would seem that the father does not provide the children with any new information. This suggests that the father he should not need to tell them that p holds when $k > 1$. But this is false. In fact, it is not hard to show that if the father does not announce p, the muddy children are never able to conclude that their foreheads are muddy (see [Fagin, Halpern, Moses, and Vardi 1995] for details).

The key point here turns out to be that the father gives the children *common knowledge* of p. I found understanding the role of common knowledge, and how Kripke structures could be used to explain what was going on in this puzzle at so many levels – logical, psychological, and linguistic – absolutely fascinating.

While puzzles were perhaps what drew me into epistemology, it was the intimate connection between epistemology and other topics of arguably more practical concern, particularly distributed computing, AI, and game theory, that kept me interested. I found it wonderful how reasoning about the knowledge of agents in a system could give insight into so many problems in all these areas, from coordination to agreement to bargaining. But here too, it was thinking about puzzles that initially brought out the connection. Aumann's famous result that we can't agree to disagree [Aumann 1976] (somewhat more precisely, it cannot common knowledge among agents that have a common prior that they have different posteriors) brought out the importance of epistemic reasoning to game theory. But this leads to an obvious puzzle: how do agents still manage to trade stocks (which, after all, involves agreeing to disagree about the expected value of a stock, which is impossible if agents cannot have different posteriors). (Of course, an agent that needs to sell a stock to finance his child's university education may be willing to trade even if he agrees that the stock that he is selling will go up relative to the rest of the market, but there is still an issue as to how agents only interested in making a profit are able to trade.) This seemed to me a great area in which to do research.

As a computer scientist, even more influential was what is perhaps my second-favorite puzzle, the *coordinated attack problem*. This problem was originally introduced by Gray [1978] as an abstraction of database issues, and can be described informally as follows (the description is taken from [Halpern and Moses 1990]:

> Two divisions of an army, each commanded by a general, are camped on two hilltops overlooking a valley. In the valley awaits the enemy. It is clear that if both divisions attack the enemy simultaneously they will win the battle, while if only one division attacks it will be defeated. As a result, neither general will attack unless he is absolutely sure that the other will attack with him. In particular, a general will not attack if he receives no messages. The commanding general of the first division wishes to coordinate a simultaneous attack (at some time the next day). The generals can communicate only by means of messengers. Normally, it takes a messenger one hour to get from one encampment to the other. However, it is possible that he will get lost in the dark or, worse yet, be captured by the enemy. Fortunately, on this particular night, everything goes smoothly. How long will it take them

to coordinate an attack?

Suppose that the messenger sent by General A makes it to General B with a message saying "Let's attack at dawn". Will general B attack? Of course not, since General A does not know he got the message, and thus may not attack. So General B sends the messenger back with an acknowledgement. Suppose that the messenger makes it. Will General A attack? No, because now General B does not know that General A got the message, so General B thinks General A may think that he (B) didn't get the original message, and thus not attack. So A sends the messenger back with an acknowledgement. But of course, this is not enough either.

What is going on here is that each time an acknowledgement is received, the level of knowledge increases by one. However, the generals never achieve common knowledge of the message. As Yoram Moses and I showed [Halpern and Moses 1990], common knowledge is required for coordinated attack (and, more generally, for coordination of all types). Moreover, in a system where communication is not guaranteed (roughly speaking, one where there is always a possibility that a message sent might not arrive), agents can never attain common knowledge of a fact that wasn't common knowledge to start with. It easily follows that coordinated attack is impossible in systems where communication is not guaranteed.

The fact that reasoning about knowledge could provide such insights was to me very exciting.

What do you see as being your main contributions to epistemology?

I started working on epistemic logic in 1983 or so, with Yoram Moses, who was then my Ph.D. student. (It's hard to believe that it was really 25 years ago!) I found much of the work in the philosophy literature in epistemology at the time quite frustrating. The focus seemed to be on finding the "right, true properties" of knowledge. The implicit picture seemed to be that we all, at heart, understood what "knowledge" was; it was the philosopher's job to explicate the notion clearly. So there would be debates about whether, for example, it was "really" the case that knowledge satisfied S4; was it really true that if you knew something, you knew that you knew it? Much of the debate was relatively informal. In some of the papers, it seemed that the definition of knowledge changed from paragraph to paragraph. (Perhaps fortunately, I've forgotten exactly which papers in this literature I found particularly annoying!) It seemed to me that there was no one "right, true" notion of knowledge. People use the word in many different ways. In any case, my interests were more pragmatic.[2] Although philosophers going back to Hintikka had argued that S5 was not the appropriate logic of knowledge (in particular, they argued that the negative introspection axiom—if you know something you know that you don't know it—was inappropriate), S5 seemed to me the most natural epistemic logic for distributed computing applications. Game theorists independently adopted S5 for their applications, with much the same motivation. While I was willing to back off from S5 for computational reasons (see below), using it gave a great deal of insight, so I was more than happy to adopt it.

The philosophy literature also focused on the single-agent case—when debating the "right, true properties" of knowledge, there's no need to consider multiple agents. (Although, to be fair, Lewis's seminal work on convention [Lewis 1969] introduced the notion of common knowledge and pointed out its importance to characterizing conventions.) Distributed systems are composed of many agents, so I was interested in the describing the knowledge of not just of one agent, but of groups of agents, and how that could change as a result of communication. Particularly important was what one agents knew about other agents' knowledge.

I see my biggest contribution as the work I did with Yoram Moses [Halpern and Moses 1990] showing how knowledge could be used to help understand and analyze distributed systems, and in defining a concrete model of knowledge in distributed systems (the so-called *runs-and-systems model*, which is developed further in [Halpern and Fagin 1989] and [Fagin, Halpern, Moses, and Vardi 1995]). "Possible worlds" and "accessibility relations" had a concrete meaning in this framework, one that computer scientists interested in designing systems could understand, even if they were not so concerned with the philosophical issues. This word introduced the vocabulary of epistemic logic to distributed computing, and emphasized the connection between common knowledge and

---

[2] While it's true that my primary interests were and continue to be somewhat more pragmatic, I must confess that I actually have a recent paper that considers definitions of knowledge in terms of (true) belief [Halpern, Samet, and Segev 2008].

important notions like agreement and coordination. This initial work was followed by a great deal of work that expanded on these themes, joint with Ron Fagin, Yoram Moses, Mark Tuttle, Moshe Vardi, Lenore Zuck, and others (largely summarized in [Fagin, Halpern, Moses, and Vardi 1995]).

Distributed computing concerns also motivated a second line of work that is enjoying newfound life in game theory. Hintikka [1962] had already observed that the standard possible-worlds semantics of epistemic logic suffers from the *logical omniscience problem*: agents knew all tautologies and that agents knew the logical consequences of their knowledge. This clearly is not a particularly accurate description of people's knowledge. Ron Fagin and I [1988] suggested that one way of dealing with the problem was in terms of *awareness*. The idea is that there is a distinction between an agent's *implicit knowledge* and her *explicit knowledge*. Implicit knowledge is defined in the usual way, as truth in all worlds; to have explicit knowledge of a fact ϕ, you must implicitly know ϕ and be aware of it. What does it mean to be aware of ϕ? That depends. It could mean that there are basic concepts in ϕ that the agent is not aware of (for example, an agent who has never heard of astrology might not be aware that the moon is in the seventh house); this is the interpretation that seems to be the one that game theorists are now focusing on. However, it might also mean that the agent cannot compute the truth of ϕ. In any case, there has been a great deal of work on awareness recently, largely published in economics journals (see, for example, [Dekel, Lipman, and Rustichini 1998; Feinberg 2004; Halpern 2001a; Halpern and Rêgo 2006a; Halpern and Rêgo 2006b; Heifetz, Meier, and Schipper 2006; Heifetz, Meier, and Schipper 2007; Modica and Rustichini 1994; Modica and Rustichini 1999] but this is just the tip of the iceberg).

What do you think is the proper role of epistemology in relation to other areas of philosophy and other academic disciplines?

Since I'm not a philosopher by training or background, I can't comment on what the proper role of epistemology is in relation to other areas of philosophy. I feel somewhat more confident in commenting on the role of knowledge in computer science and economics. We make decisions on the basis of our knowledge and belief. Understanding the connection between knowledge, belief, and action plays a key role in both computer science and economics. This connection may involve counterfactual beliefs as well (see, for example, [Halpern 2001b; Halpern and Moses 2004]). I believe that epistemology can help elucidate these issues.

What do you consider to be the most neglected topics and/or contributions in contemporary epistemology?

I'm not sure what the most neglected topics or contributions in epistemology are (I've probably neglected them along with everyone else!), so let me change the question slightly to one that I think is more interesting: What are the most pressing issues in epistemology today? To me, perhaps the most pressing issue is that of getting a good model of knowledge that takes resource-bounded reasoners, and particularly the computation involved in computing what you know, into account. This is hardly a new topic, but I think the problem is far from solved. A "good" model is one what is easy to work with, mathematically well founded, and leads to new insights. In particular, I would hope that such a model would give insights into cryptographic notions like *zero knowledge proofs* [Goldwasser, Micali, and Rackoff 1989]), and be used to define solutions concepts that take computation into account in a serious way. I believe that the notion of awareness should help here, as will perhaps the notion of *algorithmic knowledge* [Fagin, Halpern, Moses and Vardi 1995]. However, much more remains to be done.

In addition, I would like to see more work using epistemic logic on analyzing and synthesizing programs from knowledge-based specifications. The *synthesis* problem in computer science is that of deriving a program, given a specification that the program should satisfy. In many cases, what the system is supposed to do is best expressed in terms of knowledge. This is particularly true for specifications involving security; for example, we may not want an adversary to *know* certain secret information. (See [Halpern and O'Neill 2002; Halpern and O'Neill] for some discussion of how security specifications can be expressed in terms of knowledge.) There has been some work on automatically synthesizing programs from the specifications that they must satisfy (see [Bickford, Constable, Halpern, and Petride 2005; Engelhardt, Meyden, and Moses 1998; Engelhardt, Meyden, and Moses 2001]), but I think that much more can and should be done.

The first topic I mentioned, finding a good computational notion of knowledge, should have some important philosophical implications. Among other things, it would give a notion of knowledge that does not suffer from the logical omniscience problem. To the extent that it can be used to help clarify important issues in distributed computing and game theory, it will also capture important features of human epistemic reasoning. The second topic is perhaps of more pragmatic rather than philosophical interest, but it would show the applicability of epistemic reasoning to an important class of problems.

What do you think the future of epistemology will (or should) hold?

I believe that the connections between epistemology, computer science, and game theory will continue to broaden and deepen. Issues of knowledge, belief, awareness (or lack of it), computation, and counterfactuals will play an increasingly important role in the future. Philosophers could have a significant impact here and, indeed, some already have. For example, Bob Stalnaker's recent work on game theory has been quite influential (see, for example, [Stalnaker 1996]); his early work on counterfactuals [Stalnaker 1968] has turned out to be quite relevant to game theory as well. However, having an influence will require understanding, not just the philosophical issues, but the concerns of computer scientists and economists.

References

Aumann, R. J. (1976). Agreeing to disagree. *Annals of Statistics* *4*(6), 1236–1239.

Barwise, J. (1981). Scenes and other situations. *Journal of Philosophy* *78*(7), 369–397.

Bickford, M., R. L. Constable, J. Y. Halpern, and S. Petride (2005). Knowledge-based synthesis of distributed systems using event structures. In *Proc. 11th Int. Conf. on Logic for Programming, Artificial Intelligence, and Reasoning (LPAR 2004)*, Lecture Notes in Computer Science, vol. 3452, pp. 449–465. Springer-Verlag.

Dekel, E., B. Lipman, and A. Rustichini (1998). Standard state-space models preclude unawareness. *Econometrica 66*, 159–173.

Engelhardt, K., R. van der Meyden, and Y. Moses (1998). A program refinement framework supporting reasoning about knowledge and time. In J. Tiuryn (Ed.), *Proc. Foundations of Software Science and Computation Structures (FOSSACS 2000)*, pp. 114–129. Berlin/New York: Springer-Verlag.

Engelhardt, K., R. van der Meyden, and Y. Moses (2001). A refinement theory that supports reasoning about knowledge and time for synchronous agents. In *Proc. International Conference on Logic for Programming, Artificial Intelligence, and Reasoning*, pp. 125–141. Berlin/New York: Springer-Verlag.

Fagin, R. and J. Y. Halpern (1988). Belief, awareness, and limited reasoning. *Artificial Intelligence 34*, 39–76.

Fagin, R., J. Y. Halpern, Y. Moses, and M. Y. Vardi (1995). *Reasoning About Knowledge*. Cambridge, Mass.: MIT Press. A slightly revised paperback version was published in 2003.

Feinberg, Y. (2004). Subjective reasoning—games with unawareness. Technical Report Resarch Paper Series #1875, Stanford Graduate School of Business.

Gamow, G. and M. Stern (1958). *Puzzle Math*. New York: Viking Press.

Gardner, M. (1984). *Puzzles From Other Worlds*. New York:Viking Press.

Goldwasser, S., S. Micali, and C. Rackoff (1989). The knowledge complexity of interactive proof systems. *SIAM Journal on Computing 18*(1), 186–208.

Gray, J. (1978). Notes on database operating systems. In R. Bayer, R. M. Graham, and G. Seegmuller (Eds.), *Operating Systems: An Advanced Course*, Lecture Notes in Computer Science, Volume 66. Berlin/New York: Springer-Verlag. Also appears as IBM Research Report RJ 2188, 1978.

Halpern, J. Y. (2001a). Alternative semantics for unawareness. *Games and Economic Behavior 37*, 321–339.

Halpern, J. Y. (2001b). Substantive rationality and backward induction. *Games and Economic Behavior 37*, 425–435.

Halpern, J. Y. and R. Fagin (1989). Modelling knowledge and action in distributed systems. *Distributed Computing 3*(4), 159–179. A preliminary version appeared in *Proc. 4th ACM Symposium on*

Principles of Distributed Computing, 1985, with the title "A formal model of knowledge, action, and communication in distributed systems: preliminary report".

Halpern, J. Y. and Y. Moses (1990). Knowledge and common knowledge in a distributed environment. *Journal of the ACM 37*(3), 549–587. A preliminary version appeared in *Proc. 3rd ACM Symposium on Principles of Distributed Computing*, 1984.

Halpern, J. Y. and Y. Moses (2004). Using counterfactuals in knowledge-based programming. *Distributed Computing 17*(2), 91–106.

Halpern, J. Y. and K. O'Neill. Anonymity and information hiding in multiagent systems. *Journal and Computer Security 13*(3), 483–514.

Halpern, J. Y. and K. O'Neill (2002). Secrecy in multiagent systems. In *Proc. 15th IEEE Computer Security Foundations Workshop*, pp. 32–46. To appear, *ACM Transactions on Information and System Security*.

Halpern, J. Y. and L. C. Rêgo (2006a). Extensive games with possibly unaware players. In *Proc. Fifth International Joint Conference on Autonomous Agents and Multiagent Systems*, pp. 744–751. Full version available at arxiv.org/abs/0704.2014.

Halpern, J. Y. and L. C. Rêgo (2006b). Reasoning about knowledge of unawareness. In *Principles of Knowledge Representation and Reasoning: Proc. Tenth International Conference (KR '06)*, pp. 6–13. Full version available at arxiv.org/cs.LO/0603020.

Halpern, J. Y., D. Samet, and E. Segev (2008). On definability in multimodal logic II. Defining knowledge in terms of belief. Unpublished manuscript, available at www.cs.cornell.edu/home/halpern/papers.

Heifetz, A., M. Meier, and B. Schipper (2006). Interactive unawareness. *Journal of Economic Theory 130*, 78–94.

Heifetz, A., M. Meier, and B. Schipper (2007). Unawareness, beliefs and games. In *Theoretical Aspects of Rationality and Knowledge: Proc. Eleventh Conference (TARK 2007)*, pp. 183–192.

Hintikka, J. (1962). *Knowledge and Belief*. Ithaca, N.Y.: Cornell University Press.

Lewis, D. (1969). *Convention, A Philosophical Study*. Cambridge, Mass.: Harvard University Press.

Littlewood, J. E. (1953). *A Mathematician's Miscellany.* London: Methuen and Co.

Modica, S. and A. Rustichini (1994). Awareness and partitional information structures. *Theory and Decision 37*, 107–124.

Modica, S. and A. Rustichini (1999). Unawareness and partitional information structures. *Games and Economic Behavior 27*(2), 265–298.

Stalnaker, R. C. (1968). A semantic analysis of conditional logic. In N. Rescher (Ed.), *Studies in Logical Theory*, pp. 98–112. Oxford University Press.

Stalnaker, R. C. (1996). Knowledge, belief and counterfactual reasoning in games. *Economics and Philosophy 12*, 133–163.

11

Sven Ove Hansson

Professor

Royal Institute of Technology, Stockholm, Sweden

---

1. Why were you initially drawn to epistemic logic?

My earliest work in philosophical logic concerned the logic of values and norms: preference logic and deontic logic. I read the standard works in epistemic logic but they did not make me enthusiastic. What drew me to epistemic logic was the emerging new area of belief change. Philosophically it is the acquisition and loss of beliefs that is interesting, rather than the mere holding of them. The theory of belief change, as admirably outlined in the famous 1985 paper by the AGM trio (Carlos Alchourrón, Peter Gärdenfors, and David Makinson) opened up new possibilities for studying the dynamics of belief.

But from the very beginning I wanted to increase the cognitive realism of belief change modelling. Of course any model – of beliefs or anything else – has to be simplified in some respects, and there are certainly purposes for which the rather high degree of idealization in the AGM model is adequate. But some of its features are so far from belief changes in real life that complementary, more realistic models are also needed.

A central feature in the original AGM model is that of a belief set, defined as a set of sentences that is closed under logical consequence. In each belief state the agent's beliefs – or more exactly the beliefs that she is committed to affirm—are assumed to be representable by a belief set. This means that whatever follows logically from your beliefs is also assumed to be something that you believe (or at least something that you are committed to believe). The belief set is a very useful idealization. By the way, it is older than AGM theory; Isaac Levi deserves most of the credit for showing its usefulness and clarifying how it should be interpreted. In my view there is no doubt that belief sets should have a prominent role in epistemic logic.

However, the problem I saw in the AGM model is that the belief set was also taken to be the set on which operations of change are performed. This struck me as unrealistic since a logically closed set will contain both beliefs that have independent justifications and beliefs that are there only because they follow from some other beliefs that have independent justification. We can refine the AGM model by performing the actual operations of change only on the beliefs that have independent justification. The set of such beliefs is called a belief base, and obviously it will not be closed under logical consequence.

Let me explain this with an example that I have used many times before. Not surprisingly, I believe that Paris is the capital of France (p). I also happen to believe that there is milk in my fridge (q). As a logical consequence of these two beliefs, I also I believe that Paris is the capital of France if and only if there is milk in my fridge ($p \leftrightarrow q$). Hence these three beliefs are all in my belief set. Clearly, p and q are also elements of the belief base. However, $p \leftrightarrow q$ is a merely derivative belief and therefore not part of the belief base.

The significance of this difference becomes evident when I open the fridge and find that there is no milk. This discovery makes me replace my belief in q with belief in its negation ($\neg q$). I cannot then, on pain of inconsistency, retain both my belief in p and my belief in $p \leftrightarrow q$. Let us first consider this belief change from the viewpoint of a model such as AGM in which changes take place on the belief set. Since both p and $p \leftrightarrow q$ are elements of the belief set, I have to choose which of them to give up. The retraction of $p \leftrightarrow q$ does not follow automatically. In AGM a selection function or entrenchment relation has to be applied, and if its properties are adequate to the situation then it will give preference to p over $p \leftrightarrow q$ so that the former but not the latter is retained.

In a belief base framework, the contraction will instead take place on the belief base that contains p but not $p \leftrightarrow q$. Clearly p will be retained, but after q has been replaced by $\neg q$, $p \leftrightarrow q$ is no longer derivable. It disappears with q, and the option of retaining it does not even arise.

This example shows that belief bases can provide us with a more realistic picture of a process of belief change, but how important is the difference? Clearly, the selection mechanism in AGM can also ensure that p is retained and $p \leftrightarrow q$ rejected. I will give two reasons why the difference is nevertheless substantial.

The first reason is that belief bases can express distinctions that cannot be expressed if we only have belief sets. For an oversimplified example, consider the two belief bases $\{a, b\}$ and $\{a, a \to b\}$. They have the same logical closure and they are therefore statically equivalent, in the sense of representing beliefs states with the same set ofÂ beliefs actually held (i.e. the same belief set). On the other hand, they are not dynamically equivalent in the sense of behaving in the same way under operations of change. If we contract by a, then we should expect the outcome to be $\{b\}$ in the first case and $\{a \to b\}$ in the second case. This difference between the belief bases $\{a, b\}$ and $\{a, a \to b\}$ cannot be represented on the level of belief sets. Since these two bases have the same logical closure, they both correspond to the same belief set.

The expressive capacity of belief bases is perhaps most clearly seen if we consider inconsistent belief bases. Since an inconsistent set implies the whole language, there is only one inconsistent belief set namely the whole language. In contrast, there are many different inconsistent belief bases. In this respect, belief bases are much more realistic; surely there are different ways to hold inconsistent beliefs.

The second reason is that the application of operations of belief change on belief sets tends to give rise to implausible properties that do not arise if the same operations are applied to a belief base. Most notably, this is true of the famous AGM postulate of recovery. Informally speaking, recovery says that if you first remove and then reintroduce one and the same belief, then you regain the original belief set.[1]

Of course, the recovery property does not follow from the logical closure of the belief set. However, it is surprisingly difficult to construct an operator of contraction that operates directly on the belief set without also satisfying recovery. Such operators tend to instead have other, starkly implausible properties. In contrast, a contraction operator that operates on the belief base, and yields as outcome a subset of the original belief base, cannot possibly satisfy recovery. (To see that, contract c from the belief base $\{a, b\&c\}$. For the contraction to be successful, $b\&c$ has to be eliminated. It will not be regained if we afterwards reinsert c into the belief base.)

---

[1] In formal language: $K = \text{Cn}((K \div p) \cup \{p\})$. Here K is a consistent belief set, $\div$ an operator of contraction, and p a contingent belief-expressing sentence. Cn is a supraclassical consequence relation representing the logic we are operating with.

To show that recovery is not a plausible property I have used the following example: I have read in a book about Cleopatra that she had both a son and a daughter. My set of beliefs therefore contains both s and d, where s denotes that Cleopatra had a son and d that she had a daughter. I then learn from a knowledgeable friend that the book is in fact a historical novel. I therefore contract $s \vee d$ from my set of beliefs, and then I do not any longer believe that Cleopatra had a child. Soon after that, however, I learn from a reliable source that Cleopatra had a child. It seems perfectly reasonable for me to then add $s \vee d$ to my set of beliefs without also reintroducing either s or d.

To sum up, these are the types of issues that first caught my attention and made me enthusiastic about epistemic logic. I still think they are important and fruitful. Belief bases are an important step in building realistic models of the dynamic properties of beliefs. (Note that in all the examples I have given, belief sets and belief bases can be used interchangeably in static descriptions of the belief states involved. It is only when we look for representations of change that the distinction becomes important.) But of course, belief bases provide only a very rough representation of the justificatory structure of belief systems. They capture deductive justifications, but obviously many justifications of beliefs are non-deductive. Much work remains to be done to do justice to the variety of belief justifications, and to understand their role in the dynamics of belief.

Most of my early work on belief change was devoted to operations on belief bases. There are two other dominant themes in my work on belief change. First, the construction of operations other than the three standard ones (contraction, revision and expansion). Perhaps most important of these are operations in which an input sentence may or may not be accepted – a realistic feature since we do not always accept the information we are offered. Secondly, in the last few years I have spent much time on operations on finite-based belief sets. A belief set if finite-base if it is the logical closure of some finite set of sentences. Such operations turn out to be a quite interesting alternative to belief bases. They have the conceptual advantage that we do not have to divide beliefs into basic and derived, and many of the most disturbing properties of belief sets are eliminated when they are required to be finite-based. (This approach was outlined in a paper in *Erkenntnis* in 2008.)

2. What example(s) from your work, or work of others, illustrates the relevance of epistemic logic?

In answer to this I would like to say something about the application of belief revision theory in the philosophy of science. Unfortunately, studies of belief revision and studies of the development of scientific theories have been conducted without much contact, but recently some connections have been established. Erik J. Olsson and Sebastian Enqvist are just about to publish a collection summarizing these recent developments, *Belief Revision Meets Philosophy of Science*.

As I see it, the reason why there has been so little contact between epistemic logic and mainstream philosophy of science is that most frameworks developed by epistemic logicians have been unsuitable for describing the development of scientific knowledge. Belief change theory has been developed to suit two major application areas: changes in the beliefs of a single individual and changes in a computerized database. There are at least five important differences between these applications and the development of scientific knowledge. These are differences that we have to take into account when building models of scientific change.

First, the process of scientific change is collective. Decisions to accept a new idea in science are made in a fairly complex collective process. But this is not one of the collective processes studied in social choice theory. Instead it is an informal decision process that proceeds by consensus or near-consensus among the respected experts in the respective area. There is a well-developed division of labour, so that different experts are involved in different parts of science. Belief change theory has mostly been devoted to changes in individual beliefs, and the collective processes that have been studied have mostly been quite different from those at work in science.

Secondly, in scientific change the distinction between data and theory is essential. Science is primarily driven by empirical data, but the most important objects of change are scientific theories. A belief change model that can make this distinction will be able to capture important aspects of scientific change that are lost in models that cannot harbour the distinction.

Thirdly, the assimilation of new scientific data is largely an accumulative process, since data are added but seldom retracted. (There are rare exceptions such as when an observation report turns out to be incorrect.) However, this (near-)accumulativity does not apply to theoretical statements. The acquisition of new data can induce us to give up previous theoretical beliefs. This difference is an important feature of the scientific knowledge process that is not represented in standard accounts of belief change.

Fourthly, inconsistency has a much smaller role in scientific change than in common models of belief change. Strictly speaking, new data seldom contradict previous data. The crucial issue, given the accumulated data, is how these data can best be theoretically accounted for, i.e. explained. Inconsistency avoidance, the regulating force in traditional belief change, is not sufficient here. A larger role is played by a positive goal, namely search for the best explanation of the available data.

The *fifth* difference concerns contraction. Contraction is an operation in which a specified sentence is removed from the belief state but nothing new is added to it. Contraction has a central role in belief change theory, but it does not seem to have any role at all in the dynamics of scientific knowledge. We expect changes in science to be driven by new empirical data. When old scientific beliefs are given up – be they empirical or theoretical beliefs – this is not due to an external instruction as when a computer receives instructions to remove an item from its database. Instead it is (at least on the idealized level on which we need to operate) a reaction to new data that lead us to exclude old beliefs. An operation in which the addition of new information leads to the rejection of old beliefs is not contraction but revision. Contraction, in the sense that this term has in belief change theory, does not seem to have much of a role in models of scientific change.

Due to these differences I believe that models of scientific change have to differ in important ways from traditional models of belief change. As a starting-point for delineating the crucial difference, it should be noted that processes of belief change do not only include the assimilation of new information but also operations of retrieval, i.e. extraction of information from the system. In traditional models, all the hard work is done by the assimilation operation. ("Given this input, construct the new belief set!") In contrast, the subsequent retrieval operation is trivial. ("Find out what are the elements of the new belief set!") In building a model of scientific change, it may be useful to transfer much of the real work from the operation of assimilation to that of retrieval. In other words, much of the change in theory will not take place automatically upon the receipt and assimilation of new information. Instead, it will be triggered by independent processes that are more akin to the retrieval of information. This requires that we move the selection mechanism (or part of it) from the (near-accumulative) assimilation of new data to separate processes in which theories are adopted, rejected, or modified.

3. What is the proper role of epistemic logic in relation to other disciplines, for instance mainstream epistemology, game theory, computer sciences or linguistics?

The role of formal methods in philosophy has often been either drastically over- or understated. On one side we have the Scylla of hegemonical contentions. Bertrand Russell, whose life achievement I deeply admire, unfortunately made some exaggerated claims on what logic can do, claiming for instance that "every philosophical problem, when it is subjected to the necessary analysis and purification, is found either to be not really philosophical at all, or else to be, in the sense in which we are using the word, logical". (*Our knowledge of the external world*, p. 14) This is claiming too much.

On the other side we have the Charybdis of capitulation. Some philosophers regard logic as of no help at all in elucidating philosophical issues. Even von Wright, who himself invented whole new areas of philosophical logic, was pessimistic about the future prospects of formal philosophy. (See his essay on logic in Frederick Stoutland's recent volume *Philosophical Probings: von Wright's Later Work*.) But as I hope to already have illustrated, this is an untenable position.

We have to steer between these two extremes. What logic can do is to provide us with useful formal models of philosophical subject-matter, for instance epistemic processes. Like all other model-building exercises, the use of logic in epistemology involves the streamlining and simplification of concepts and ideas. In other words it is an idealizing process. Therefore it always involves a trade-off between simplicity and faithfulness to the original.

Once we have realized that this is the role of logic we will also see that there are no a priori grounds why logical languages should be better suited than other symbolic languages to model epistemic processes. The usefulness of logic cannot be taken for granted a priori; it has to be proved by the construction of useful logical models.

Different types and styles of models of one and the same subject-matter should be seen as complements rather than as competitors. Probabilistic models are immensely useful in epistemology, not least since they capture other features of belief and knowledge than what logic usually does. Our beliefs seem to vacillate between the yes-or-no properties captured in truth-functional logic and the graduality captured in probability theory. We still lack an overarching formal theory that can make sense of this vacillation.

There are additional types of formal models that are useful in epistemology. Some collective epistemic processes are best captured in game-theoretical models. In order to describe the influence of epistemic (and other) values on beliefs, decision-theoretical models will be useful. We need a variety of models of epistemic processes. Logic does not have a monopoly.

4. Which topics and/or contributions should have had more attention in late 20th century epistemic logic?

Generally speaking, I believe that we have paid too little attention to applications other than those in computer science. There are at least three application areas that would have been worth more attention.

First: Philosophy of science. I have already mentioned the need for formal models of the development of scientific knowledge. But this is not the only issue in the philosophy of science that would gain from more inputs from epistemic logic. The study of epistemic values is one example. Yet another is the study of explanatory reasoning in science.

Secondly: Legal reasoning. Due to its focus on argumentation and inference, legal reasoning is well suited for formal modelling. The study of evidence has occasionally made use of formal methods, but it remains to investigate what can be achieved with the tools of modern epistemic logic in this area.

Thirdly: The mechanisms of collective epistemic failure. Psychologists have studied processes that lead a whole group to make the same epistemic mistake, such as groupthink and prejudice. Logical studies of collective belief revision processes have mostly dealt with successful practices, but the tools of epistemic logic are well suited to investigate epistemic failures as well.

5. What are the most important open problems in epistemic logic and what are the prospects for progress?

I must again emphasize the relationship between probabilistic and non-probabilistic models of belief. Another important, very general field is reasoning under cognitive limitations. How can cognitive limitations be realistically represented? How can we model systems of reasoning and belief that operate under such limitations?

Let me also mention the issue of compartments of mind and contained inconsistencies. Intuitively, we can think of a person's belief system as consisting of several (partly overlapping) compartments of mind. Inconsistencies within one such compartment, or between elements of two or a few compartments, can be kept contained and do not make the whole belief system inconsistent. (It seems to be quite feasible to believe both that Jesus was a human being and that Jesus was not a human being, without believing that the moon is made of cheese.) The logical features of such containment have only begun to be investigated.

Finally, the influences of values on beliefs are well suited for logical investigations, not least since we now have well-developed logical theories both for belief change and for changes in values (preferences). This concerns some of the central issues in philosophy, such as the is-ought and fact-value issues and the relationship between practical and theoretical philosophy. Recently, sufficiently sophisticated logical tools to investigate these problem-areas have become available. But the work remains to be done.

12

Aviad Heifetz

Professor

The Economics and Management Department

Open University of Israel, Israel

---

1. Why were you initially drawn to epistemic logic?

I feel compelled to start by saying that there is something peculiar in this list of 5 questions. The list gives the impression that epistemic logic is some kind of discipline standing in its own right, independently among other intellectual or scientific disciplines. But in fact, to my mind the state of affairs is just the opposite: epistemic logic is first and foremost a *language* to express what individuals can know or believe about the world, *y compris* the knowledge and beliefs of others. So just as we don't put on an equal footing and ask similar questions about, say, Physics and French, be they as beautiful and complicated and engaging as they both are, any question we can ask about a language like epistemic logic has a *fortiori* a self-referential, ars *poetic* type of nature.

The first question here does not evade this nature. It is meaningless to ask a child 'why were you initially drawn to your mother tongue?' This question is meaningless even though it is grammatically flawless. This is so because the child's perception and notion of her own self, as an enduring entity distinguished from others and from the physical world, was actually continuously forged during the process of acquiring her mother tongue and using it in communication with others. Asking 'Why were you...' assumes a decision maker who made a conscious choice at some specific point in time. But the child did not exist as a conscious decision maker before she was already immersed in her mother tongue. This is why she cannot even reflect in retrospect about her experience of learning to speak.

Epistemic logic is definitely not my mother tongue, so I *am* able to reflect back about the years in which I gradually began to employ this language; the analogy I alluded to above is nevertheless relevant, because I never had an intention to indulge myself in epistemic logic. In fact, just the opposite was true. After a first inquiry into probabilistic type spaces in my M.Sc. thesis, I tried to pursue some game theoretic questions for my Ph.D.. Having trouble overcoming them, I found the time to just 'take off the table' a few questions that emerged from my M.Sc. thesis. These led to a completely different Ph.D. than the one I first intended to write, and speaking a new language I did not even know to name at that time as 'epistemic logic'.

My very first encounter with this language occurred when I took a graduate course on Information, Probability and Games with Sergiu Hart at Tel Aviv University in 1987 or 1988. One meeting was dedicated to the Mertens-Zamir (1985) construction. Sergiu described it as mathematically challenging, and indeed it was – I understood little of its math, but I was fascinated by its aesthetic elegance. Shortly afterwards, I heard Dov Samet presenting his joint paper with Dov Monderer "Approximating Common Knowledge with Common Belief (1989), in two consecutive sessions of the Game Theory and Mathematical Economics seminar at Tel Aviv University. Afterwards I approached Dov Samet with some 'brilliant' idea I had in mind. I have no recollection what this nonsense was about, but luckily enough Dov was patient, and referred me to read the Parikh and Krasucki (1990) paper "Communication, Consensus and Knowledge, which was still not published at the time. Their claim that consensus can be reached without this being common knowledge was intuitively appealing, but the puzzle *was* that within the formal partition model that they were using, consensus *was* accompanied by common knowledge.

So I set out to write what eventually became my paper "Comment on Consensus without Common Knowledge (1996) (but was first phrased in a much more cumbersome framework than needed ...) . I was fortunate to do so, because when I attended my first international conference, the 1991 Game Theory festival at Stony Brook, I had the guts to approach Herakles Polemarchakis and tell him that I have a work relating his paper with John Geanankoplos *We can't disagree forever* (1982) and the Parikh-Krasucki paper. His reaction was: "Good! I never understood the Parikh-Krasucki paper - maybe you would be able to explain it to me."

So I tried, and consequently Herakles invited me for a 10-day visit to CORE in Universite Catholique de Louvain, Belgium, where I could meet and discuss not only with Herakles himself, but also with Jean Francois Mertens, whose work with Shmuel Zamir on the universal type space I had by then extended in my M.Sc. thesis paper (Heifetz 1993) to a more general setting. In that visit I also met Francoise Forges and Enrico Minelli, with whom I later collaborated in several projects. I recall being naively frustrated at the time that this particular short visit didn't lead to any new paper, but as always Herakles was generous enough to invite me to CORE several more times. In a subsequent visit I met Luc Lismont and also Philippe Mongin, with whom I had a fruitful collaboration in the sequel on *Probability Logic for Type Spaces* (Heifetz and Mongin 2001), and even more importantly, who asked me many of the questions that eventually led me to formulate my ideas not only in semantic models but also within syntactic models of epistemic logic. These led also to a study of Bob Aumann's working paper on interactive epistemology, which was eventually published in two parts (Aumann 1999a,b), and subsequently also to an on-going direct exchange of ideas with Aumann (see Aumann and Heifetz 2002).

This way I began to speak a language that I never intended to learn, but had to admit that I'm speaking once Francoise Forges suggested that I wrap up the papers that I wrote that far as my Ph.D. thesis, and apply for a post-doc year at CORE. The interaction I had with all these people during this intensive post-doc year led to even further interest and work on models of interactive epistemology. I'm happy to acknowledge and thank all the generous people mentioned above for being my intellectual mothers and fathers, who immersed me into this language.

2. What example(s) from your work, or work of others, illustrates the relevance of epistemic logic?

I can only faithfully testify to the importance of epistemic logic in game theory. Though this is not readily admitted, the very notion of a strategy in an extensive-form game actually involves beliefs: In each information set of a player, her strategy specifies not only her current action, but also what she is **certain** she will be doing along each potential continuation path of the game, and what she would have done were the game to evolve differently than it has already evolved. (Rubinstein 1991 makes a related point.) Thus, every solution concept in game theory that pins down a subset of the players' strategies inevitably also points at some beliefs and mutual beliefs. Typically, these beliefs remain implicit in the specification of the solution concept, and oftentimes this is a source of confusion.

Here is where epistemic game theory becomes important. Consider the notion of backward induction and subgame perfection, introduced by Selten. Selten's suggestion was to analyze the game as if it were, effectively, a dynamic optimization problem of a single agent applying the principle of optimality. Selten's original motivation was to introduce a solution concept that voids empty, non-credible threats, e.g. in two-stage games in which every player gets to play only once. In view of its effectiveness in such games, his ingenious and elegant suggestion has rightfully gained immanence as the basic solution concept for extensive-form games. However, when applied to longer games, in which some players are called to play several times, the subgame perfection notion becomes more strained: if a player has already deviated from the backward induction path, why would a player at a consecutive stage plan her moves *assuming* the first player would stick to his subgame perfect strategy, from which he has already deviated? (see e.g. Samet 2005)

Epistemic game theory has studied this tension at length. On one side, elaborate studies (Aumann 1995, Asheim 2002, Asheim and Perea 2005; see also the extensive survey in Perea 2007) make explicit the assumptions on the optimistic, relentless belief in future rationality (irrespective of past behavior) implicit in backward induction per se, and how weakening these assumptions, e.g. by assuming only initial belief in rationality (Ben-Porath 1997) can lead to different outcomes.

A more subtle argument for the backward induction outcome arrives, though, from a completely different direction. Pearce's (1984) notion of *extensive-form rationalizability* embodies the complementary idea of *forward induction*, namely that at each information set, each player should try to rationalize as much as possible the opponent's past behavior (i.e. attribute to her the highest possible degree of rationality, belief in her opponent's rationality, belief in *his* belief in rationality etc.), and use this best rationalization for assessing her likely future behavior. Pearce's notion of extensive-form rationalizability was given an epistemic characterization by Battigalli and Siniscalchi (2002) using the notion of rationality and correct common *strong* belief in rationality.

On the face of it, this important forward-induction notion has little to do with backward-induction, and indeed, it is generally the case that extensive-form rationalizable strategies differ from the backward-induction strategies of the players. Surprisingly, however, in generic perfect-information extensive-form games, the backward induction path coincides with the path induced by any extensive-form rationalizable strategy profiles (Reny 1992, Battigalli 1997). This suggests that we are rightly attracted to the backward induction solution, but possibly for the 'wrong' reasons: we apply a dynamic optimization algorithm that is conceptually unsuitable for an *interactive* situation, while the true justification lies with a strategic, forward-induction notion, whose nature is exposed to us by an epistemic logic analysis.

Epistemic game theory has thus been applied to provide an epistemic characterization of various solution concepts – Nash equilibrium (Aumann and Brandenburger 1995), rationalizability (Tan and Werlang 1988), common p-belief in rationality (Hu 2007) and admissibility (Brandenburger Friedenberg and Kiesler 2008). It has also been used to define new solution concepts: in Ben-Porath and Heifetz (2010) we define CKRMC prices as the prices that are consistent with common knowledge of rationality and market clearing in an economy with asymmetric information. We then (generically) characterize it using a heuristic procedure, and it is illuminating to realize that in some (non-generic) games the heuristic actually *fails* to capture the precise epistemic notion. Without an epistemic model, we would have stopped short of expressing the precise content of this solution concept.

In epistemic game theory the epistemic models are *type spaces*, in which each type of each player is defined by its belief – a sigma additive probability measure - about the fundamentals of the game (the "states of nature") and about the other players' types. The idea of a type space was the path-breaking contribution of Harsanyi (1967-68), designed to capture beliefs and mutual beliefs: The belief of a player's type about the types of her opponents defines, in particular, a belief of hers about their beliefs about nature; since this is true for the opponents' types as well, the player's belief about the opponents' types defines also her belief about their beliefs about others' beliefs about nature, and so forth.

Is any such hierarchy of beliefs (about nature, about others' beliefs about nature etc.) entertained by some type in some type space? Under suitable topological assumptions the answer is positive (Mertens and Zamir 1985, Brandenburger and Dekel 1993, Heifetz 1993, Mertens, Sorin and Zamir 1994), and the space of all these belief hierarchies constitutes the *universal type space* –

into which every other type space (with the same space of states of nature) can be mapped by a belief-preserving morphism. Moreover, this universal space is complete – each potential belief of a player about nature and the other players' types is maintained by exactly one of the player's types in the universal space, and this space is also the canonical space of maximally-consistent sets of formulas in an infinitary probability logic (Meier, forthcoming) which extends the sound and complete finitary probability-logic of type spaces (Heifetz and Mongin 1995). Even in the absence of these topological assumptions (guaranteeing the regularity of probability measures and the applicability of a Kolmogorov extension theorem)) a universal space exists (Heifetz and Samet 1998b), but some hierarchies of beliefs might be maintained by no type in no type space (Heifetz and Samet 1999).

The state of affairs is more complicated for the case of knowledge. In a seminal and immensely influential contribution, Aumann (1976) introduced knowledge spaces to game theory. In a knowledge space each player has a partition on the set of states of the world, and a knowledge space is thus a multi-player partitional Kripke structure. Aumann (1999a) showed, essentially, that the canonical model of multi-player S5 maximally-consistent sets of formulas is a knowledge space, and hence manifests the soundness and completeness of multi-player S5 logic with respect to knowledge spaces; Fagin et al. (1991) (see also Fagin et al. 1995) carried out a similar enterprise from a computer science perspective, also for weaker epistemic logics. However, unlike in the case of (sigma-additive) probabilistic belief, this canonical model is not universal; in fact, there is no universal knowledge space (Heifetz and Samet 1998a; i.e., there is no terminal object in the category of knowledge spaces with knowledge-preserving morphisms). Equivalently, the canonical model (of maximally-consistent sets of formulas) of no infinitary-logic extension of multi-player S5, no matter what cardinality of disjunctions and conjunctions it allows, would constitute a universal knowledge space (Heifetz 1997).

I will not be able to do justice here to the exponentially growing literature that employs epistemic concepts in game theory, and hence relies – directly or indirectly - on their logical foundations. The prevalent assumption of a common prior in type spaces has definitely received such a foundational inquiry (Morris 1994, Halpern 2002, Feinberg 2000, Heifetz 2006); correlations that can arise in strategic behavior receive an epistemological analysis (Ely and Peski 2006, Dekel, Fudenberg and Morris 2007) – and these are just two topics out of many. My own current fascination is with my on-going research project with Martin Meier and Burkhard Schipper on modeling and solving strategic interaction involving unawareness (Heifetz, Meier and Schipper 2006, 2008, 2010).

3. What is the proper role of epistemic logic in relation to other disciplines, for instance mainstream epistemology, game theory, computer sciences or linguistics?

I have partly replied to this question at the beginning. As a language, epistemic logic provides a vocabulary to communicate and express ideas about interaction and mutual knowledge and belief between individuals, be they human beings in the society, countries in the international arena, computers in a distributed system, or one's own ego, super-ego and id. A new exercise, a new definition or a new axiomatization in epistemic logic are only relevant if they are somehow inspired and related to such interactions. Exercises that relate themselves solely to previous ones and constitute incremental developments of them are much less exciting as such. As is the case for any normal science (in the sense of Kuhn 1962), we cannot avoid finding ourselves indulging in such puzzle solving most of our time. But we should better not forget also the wider picture, which gives the taste to it all.

4. Which topics and/or contributions should have had more attention in late 20th century epistemic logic?

Extensive-form rationalizability (Pearce 1984, Battigalli 1997), mentioned above, has definitely received much less attention than it deserves. Only recently do game theorists gradually realize the far-reaching implications of assuming common knowledge (or common p-belief) of rationality, and hence do not restrict any more their focus solely to equilibrium concepts. Expanding this type of analysis to dynamic interaction is vital, and Pearce's extensive-form rationalizability is an essential and impressive primary step in this direction. It would be fascinating to see how further application of epistemic logic to dynamic interaction would evolve.

5. What are the most important open problems in epistemic logic and what are the prospects for progress?

There is a sense in which the combination 'epistemic logic' is a contradiction in terms. Logic is about analysis and deduction, about systematic method. Knowledge about the world, in contrast, is more often than not an event and an active encounter rather than a static state of mind that can be systematically attained and maintained (Gadamer 1965). The 'Aha!..." experience of revealing something new is an occurrence that can never be guaranteed by a system of derivation from premises; it can only be *facilitated* by such measures.

A different angle of this is vividly expressed in Wittgenstein's On Certainty (1969). All theorems are, literally speaking, *trivialities* (emanating from triplets of the form $A \to B$ and $B \to C$ implying $A \to C$), while anything we are genuinely sure about is non-trivial. Wittgenstein claims that our *world-picture* or *language-game* is very different from a system of axioms which are independent of one another; just the opposite – the sentences in our language-game, that we *cannot even doubt* without 'losing our mind' (like "Earth existed during the last hundred years"), all *support* one another and as if *lean one against the other*. The language-game is like a river-bed, which keeps evolving with time in shape and pattern while retaining its essence as the same river-bed. And just as the river-bed evolves with the movement of the water, the wind and the soil, the language-game evolves with the actions and interactions of human beings. As children we learned what a chair is not by a dictionary definition, but rather by the request of those who took care of us to *sit* on a chair, and via our very action of *sitting*. Similarly, what is evident for us keeps evolving not merely by scholastic argumentation, but mostly by *employing* language in new ways.

Since the enlightenment we live in a world in which rationality subsumed religion as a source of authority, and has thus also subsumed some of its religiosic qualities. But real progress is not only about what can be attained by mastering the world more rationally and more effectively, by *knowing* how to produce more with less, but also about *how* to aim at goals in a more human way, tuned to human flourishing rather than mostly to bottom-line achievements and end-of-the-day consumption (Weil 1952). The 'Aha!..." knowledge, that catches us unprepared in our excursions and direct contact with the inherently chaotic world , is no less important for this latter type of progress than the methodological knowledge attained at the concluding line of the analytic proof.

Attending to the process rather than only to the end results requires a *combination* of science and art, of the "heart of Judgment" (Thiele 2006) of Aristotlian *phronesis*. I believe it is time for epistemic logic – in the wide sense of the term – to pull itself out of its internal contradiction and to expand its attention from whatever can be said on the basis of cold-blooded calculation to what can be known about the world and about others using empathy and affect as well.

References

Asheim, G.B. (2002), On the epistemic foundation for backward induction, *Mathematical Social Sciences* 44, 121-144.

Asheim, G.B. and A. Perea (2005), Sequential and quasi-perfect rationalizability in extensive games, *Games and Economic Behavior* 53, 15-42.

Aumann, R.J. (1976), Agreeing to Disagree, *Annals of Statistics* 4,1236-1239.

Aumann, R.J. (1995), Backward induction and common knowledge of rationality, *Games and Economic Behavior* 8, 6–19.

Aumann, R.J. (1999a), Interactive epistemology I: Knowledge, *International Journal of Garne Theory* 28, 263-300.

Aumann, R.J. (1999b), Interactive epistemology II: Probability, *International Journal of Game Theory* 28, 301-314.

Aumann, R.J. and A. Brandenburger (1995), Epistemic conditions for Nash equilibrium, *Econometrica* 63, 1161-1180.

Aumann, R.J. and A. Heifetz (2002), Incomplete Information, in R.J. Aumann and S. Hart (eds.), *Handbook of Game Theory* vol. 3, 1665-1686, Elsevier/North Holland.

Battigalli, P. (1997). On rationalizability in extensive games, *Journal of Economic Theory* 74, 40-61.

Battigalli, P. and M. Siniscalchi (2002). Strong belief and forward induction reasoning, *Journal of Economic Theory* 106, 356-391.

Ben-Porath, E. (1997), Rationality, Nash equilibrium and backwards induction in perfect information games, *Review of Economic Studies* 64, 23-46.

Ben-Porath, E. and A. Heifetz (2010), Common Knowledge of Rationality and Market Clearing in Economies with Asymmetric Information, mimeo.

Brandenburger, A. and E. Dekel (1993), Hierarchies of Beliefs and Common Knowledge, *Journal of Economic Theory* 59, 189-198.

Brandenburger, A., A Friedenberg and H.J. Keisler (2008), Admissibility in Games, *Econometrica* 76, 307-352.

Dekel, E., D. Fudenberg and S. Morris (2007), Interim Correlated Rationalizability, *Theoretical Economics* 2, 15–40.

Ely J.C. and M. Peski (2006), Hierarchies of belief and interim rationalizability, *Theoretical Economics* 1, 19–65.

Fagin, R., J.Y. Halpern and M.Y. Vardi (1991), A model-theoretic analysis of knowledge, *Journal of the Association for Computing Machinery* (ACM) 91, 382-428.

Fagin, R., J.Y. Halpern, M. Moses and M.Y. Vardi (1995), *Reasoning about Knowledge*, MIT Press.

Feinberg, Y. (2000), Characterizing common priors in the form of posteriors, *Journal of Economic Theory* 91, 127-179.

Gadamer, H.G. (1965), *Truth and Method*, Seabury Press.

Geanakoplos, J. and H. Polemarchakis (1982), We can't disagree forever, *Journal of Economic Theory* 28, 192-200

Halpern, J.Y. (2002), Characterizing the Common Prior Assumption, *Journal of Economic Theory* 106, 316-355.

Harsanyi, J. (1967-8), Games with incomplete information played by 'Bayesian' players, Parts I-III, *Management Science* 8,159-182, 320-334, 486-502.

Heifetz, A. (1993), The Bayesian Formulation of Incomplete Information - the Non-Compact Case, *International Journal of Game Theory* 21, 329-338.

Heifetz, A. (1996), Comment on Consensus without Common Knowledge, *Journal of Economic Theory* 70, 273-277.

Heifetz, A. (1997), Infinitary S5 Epistemic Logic, *Mathematical Logic Quarterly* 43, 333-342.

Heifetz, A. (2006), The Positive Foundation of the Common Prior Assumption, *Games and Economic Behavior* 56, 105-120.

Heifetz, A., M. Meier and B. C. Schipper (2006), Interactive unawareness, *Journal of Economic Theory* 130, 78-94.

Heifetz, A., M. Meier and B. C. Schipper (2008), A Canonical Model for Interactive Unawareness, *Games and Economic Behavior* 62, 304-324.

Heifetz, A., M. Meier and B. C. Schipper (2010), Dynamic Unawareness and Rationalizable Behavior, mimeo.

Heifetz, A. and P. Mongin (2001), Probability Logic for Type Spaces, *Games and Economic Behavior* 35, 31-53

Heifetz, A. and D. Samet (1998a), Knowledge Spaces with Arbitrarily High Rank, *Games and Economic Behavior* 22, 260-273.

Heifetz, A. and D. Samet (1998b), Topology-Free Typology of Beliefs, *Journal of Economic Theory* 82, 324-341.

Heifetz, A. and D. Samet (1999), Coherent Beliefs are not Always Types, *Journal of Mathematical Economics* 32, 475-488.

Hu, T-W (2007),On p-rationalizability and approximate common certainty of rationality, *Journal of Economic Theory* 136, 379-391.

Kuhn T. (1962), *The Structure of Scientific Revolutions*, University of Chicago Press.

Meier, M. (forthcoming), An infinitary probability logic for type spaces, *Israel Journal of Mathematics*.

Mertens, J.-E, S. Sorin and S. Zamir (1994), *Repeated Games*, Discussion Papers 9420, 9421, 9422, Center for Operations Research and Econometrics (CORE), Université Catholique de Louvain.

Mertens, L-F., and S. Zamir (1985), Formulation of Bayesian analysis for games with incomplete information, *International Journal of Game Theory* 14, 1-29.

Monderer, D. and D. Samet (1989), Approximating Common Knowledge with Common Belief, *Games and Economic Behavior* 1, 170-190.

Morris, S. (1994), Trade with heterogeneous prior beliefs and asymmetric information, *Econometrica* 62, 1327-1347.

Parikh, R. and P. Krasucki (1990), Communication, Consensus and Knowledge, *Journal of Economic Theory* 52, 178-189.

Pearce, D.G. (1984). Rationalizable strategic behavior and the problem of perfection, *Econometrica* 52, 1029-1050.

Perea, A. (2007), Epistemic Foundations for Backward Induction: An Overview, *Interactive Logic Proceedings of the 7th Augustus de Morgan Workshop, London*. Texts in Logic and Games 1 (Johan van Benthem, Dov Gabbay, Benedikt Löwe (eds.)), Amsterdam University Press, 159-193.

Reny, P.J. (1993), Common belief and the theory of games with perfect information, *Journal of Economic Theory* 59, 257-274.

Rubinstein, A. (1991), Comments on the Interpretation of Game Theory, *Econometrica* 59, 909-924.

Samet, D. (2005), Counterfactuals in Wonderland, *Games and Economic Behavior* 51, 537–541.

Tan, T. C-C and S.R.d.C. Werlang (1988), The Bayesian foundations of solution concepts of games, *Journal of Economic Theory* 45, 370-391.

Thiele, L.P. (2006), *The Heart of Judgment: Practical Wisdom, Neuroscience, and Narrative*, Cambridge University Press.

Weil, S. (1952), *The Need for Roots*, Routledge.

Wittgenstein, L. (1969), *On Certainty*, Blackwell.

13
Jaakko Hintikka

Professor of Philosophy
Boston University, USA

In a sense, I became aware of the possibility and potentialities of epistemic logic in the same way as the philosophical community at large. Even though the detailed history remains to be written, it is clear that what is now called "philosophical logic" started from G.H. von Wright's realization (expressed in his 1952 work *An Essay in Modal Logic*) that the methods used in modal logic can be used to study the logical behavior of such philosophically important notions as obligation, knowledge, belief, etc. It appears that von Wright's inspiration came from the vivid model-theoretic ("intuitive") meaning of the technique of distributive normal forms where different disjuncts describe different kinds of "possible worlds".

I was at the time studying and later working with von Wright and thus came be automatically involved in the development of the semantics of modal logic in general and (as an aspect of that enterprise) in the development of epistemic logic in particular. Later in 1962 my book *Knowledge and Belief: An Introduction to the Logic of the Two Notions* helped to solidfy epistemic logic as an independent discipline.

My first idea — in some ways, the cornerstone of all epistemic logic — is that the notion of knowledge serves essentially to express a dichotomy between two mutually exclusive and collectively exhaustive classes of "possible worlds" (scenarios, states of affairs or courses of events). A probability theorist would probably call them sets of sample-space points or perhaps events. They are the class of the worlds compatible with everything that is known and the class of those incompatible with it. This dichotomy determines the structure of epistemic logic to a considerable extent. Among the implications and suggestions of this idea there are the following:

1. Other epistemic notions (information, belief, perception, memory etc.) likewise express analogous dichotomies. Hence their logic is similar to epistemic logic. (In many contexts, epistemic logic should really be thought of as a logic of information rather than of Knowledge in philosophers' exalted sense.)

2. If so, since one of the main factors of the definition of knowledge is to distinguish it from these other epistemic notions, the logic of knowledge has little to do with the definition of knowledge.

3. Logically speaking, the dichotomy idea means essentially that epistemic logic (w. one knowledge operator) is a Boolean algebra with an operator. Such algebras were studied by Tarski and his associates around 1950. Their work anticipates much of the later technical work on the semantics of modal logics.

Such logics are useful, for instance in handling large bodies of information. Unsurprisingly, this has prompted computer scientists, for instance data-base theorists, to take an active interest in reasoning about knowledge, in other words, epistemic logic.

4. On the face of things, such an epistemic logic is nevertheless not very promising philosophically as a logic, that is to say as a study of which knowledge statements follow from which ones. Such consequence relations depend essentially only on the logical relations between what is known. For instance, KF implies logically KG if and only if F implies G. This is essentially a conservation law for "knowing that". Much less trivially, we can also formulate similar "conservation laws" for "knowing + wh-clause"

This suggests that epistemic logic does not offer much mileage for dealing with questions of epistemic justification.

5. However, epistemic relations become of tremendous interest when we begin to study knowledge acquisition. The paradigmatic way of such acquisition is by asking questions. Hence the by far most promising branch of epistemic logic for applications is the logic of questions and answers. It is no exaggeration that a significant logic of questioning came about only via epistemic logic. This is of huge importance epistemologically, in that epistemic logic shows how all information-seeking can be modeled as a questioning process.

6. It is also seen immediately that the main epistemic operators can be construed as quantifiers over classes of possible worlds. (This of course is true also of modal notions in general.) This shows where the most intricate and most difficult purely logical problems of epistemic logic (and of modal logics in general) are to be found. As in other cases, they are at bottom questions about the interplay of different quantifiers (including epistemic and modal operators). The surprising fact is that the interplay of different quantifiers with each other and with propositional connectives is still not satisfactorily mastered by most contemporary logicians. From Frege to Kripke, logicians have failed to see that the formal dependence relations between quantifiers are the main means of expressing the all-important actual ("material") dependence relations between variables. Once this is seen, it is also seen that the traditional Frege-Russell logic of quantifiers is defective; it cannot serve to express all possible dependence patterns among variables.

These problems are magnified in epistemic and modal logics. For one thing, the range of values of a quantifier is affected by its being dependent on an epistemic operator. In modal logics, analogous phenomena give rise the notorious problems of "quantifying in" and cross-identification.

The most significant novelty of the "second generation epistemic logic" I have developed is the recognition of the role of dependence relations. They affect the very logical form of knowledge codifications. For instance, what has been called "quantifying into" the scope of an epistemic operator turns out to be quantification independent of that that epistemic operator.

It is obvious to me that the most promising field — promising both purely logically and philosophically — of future work in epistemic logic is the study of the dependence and independence relations between epistemic operators and/or quantifiers and of the analysis their mode of operation. This is the direction of my recent and ongoing thinking about epistemic logic.

14

Wiebe van der Hoek

Professor of Computer Science
University of Liverpool, UK

---

1. Why were you initially drawn to epistemic logic?

In fact originally for me it was 'just one' of the possible interpretations of modal logic. The axioms that are often assumed for the epistemic logic S5 make it a relatively simple logic, yet it is fascinating how much there is to say about epistemic logic, especially when it is combined with other modalities. For instance, combining the two intuitively simple concepts of knowledge and time gives almost immediately rise to computationally very complex systems. Also, there are many seemingly different systems and semantics for time and knowledge, or computation and knowledge, and more often than not they appear to be technically equivalent, or at least very closely related. At the moment for instance it is intriguing to see how good old-fashioned AGM belief revision neatly fits in a very general theory of dynamic doxastic logic, which in turn grew out of dynamic epistemic logic.

As a student I could not get enough of puzzles and paradoxes involving notions like common knowledge, auto-epistemic logic and only knowing. Notions that by their nature lend themselves to wide philosophical discussions and debate, but for which we find at least some concrete answers by looking at what the logical systems tell us about them.

When I now teach modal logic I find it very difficult to motivate and explain the accessibility relation on Kripke models in other terms than epistemic indistinguishability.

2. What example(s) from your work, or work of others, illustrates the relevance of epistemic logic?

There are many (as I am supposed to say) but one of them is how the standard system of multi-agent knowledge S5 has found its way in the agent literature. Where in the 1990's one had to really make an effort to sell a possible world model in the agent community, and where BDI, one of the first frameworks to do so is admittedly complex, nowadays it is well accepted to describe an agent model and to summarize the informational status of the agents by saying that they are modeled in an 'S5-like way': this concept does not need further explanation anymore.

By just introducing a knowledge operator in an existing framework, it is often the case that one realizes quickly that the framework without it (i.e., when perfect information is assumed) was in fact rather restrictive. An example of this is the notion of a knowledge program, where one takes the fact seriously that an agent can only act upon the information it has available, and hence the condition in any if-then program construct should not refer to the state of the world, but rather to the information available about it. A second but related example is found in a relatively new area of AI research called General Game Playing, where the idea is that agents, rather than becoming experts in a game like chess, are confronted with a description of a new game, and supposed to engage in a competition with others. There have been some proposals for a General Game Playing description language, and now, as soon as its principles are understood, researchers start working on adding features to deal with incomplete information.

3. What is the proper role of epistemic logic in relation to other disciplines, for instance mainstream epistemology, game theory, computer sciences or linguistics?

The role of epistemic logic, and that of logic, or formalisation in general, is that it helps us to become aware of what specific choices in the modeling of a concept lead us to. Philosophical discussions about the notions of knowledge and beliefs are interesting, but they don't lead often to somewhere if they are not formalized.

Currently there is a lot of interest in the role of epistemic logic in explaining solution concepts in games, and more generally in using (modal) logic to reason about games. Game theorists typically ask what this buys them in terms of new theorems, but the point is that (epistemic) logic in many cases gives a justification for what game theorists are doing, and often is able to lay bare hidden assumptions in a game theoretic analysis. An example of this is Aumann's theorem that common knowledge of rationality is sufficient to explain how backward induction yields a Nash equilibrium in an extensive game with pure strategies. On the face of it, this theorem is very convincing, but if one digs deeper in its formalisation and that of its proof, one finds many delicate issues.

The role of epistemic logic in relation to computer science has many faces. If one accepts that the run of a program defines a trace along which decisions are made, and that decisions typically depend on the information that is available at the point of decision making, one encounters the rich and intriguing domain of 'knowledge and time', or of 'knowledge and action'. Epistemic logic helps making assumptions on this domain to become explicit, assumptions like 'perfect recall' and 'no learning'. But at least as interesting is the fact that computer science helps answer the question as to how 'computationally complex' certain logical problems are. Philosophers have a tendency to say they don't care about the complexity of, say satisfiability in a system for knowledge and time, or how hard it is to model-check certain liveness and fairness properties. But the answer to those questions give interesting insights in how different notions interact, and how complex the resulting systems become!

4. Which topics and/or contributions should have had more attention in late 20th century epistemic logic?

Regarding a topic that could have received more attention: I have not seen many convincing (probably read: mathematically neat) theories that deal with bounded rationality. And I have seen only a very few good proposals that deal with the problem of logical omniscience, without looking a bit artificial.

When it comes to a specific contribution, there is a paper by Brafman, Halpern and Shoham from the late 1990's on the knowledge requirements of tasks, where the idea is that in order to successfully perform a task, it would be interesting to be able to quantify the complexity of the knowledge that is required to fulfill it. In the same way as computational complexity had an impact on computer science in providing a way to classify problems according to their hardness, a theory of the knowledge complexity of tasks might give AI (think about robotics and planning) a tool to quantify what kind of knowledge the problem solver needs (what are the sensors we should equip a robot with) to complete a task. I found and find this a very interesting and promising idea, but as far as I know, the paper had yet little follow up.

5. What are the most important open problems in epistemic logic and what are the prospects for progress?

It is remarkable to note that epistemic logic for computer science was originally motivated by Halpern et al. at IBM as a formalism to reason about communication protocols. In the meantime, epistemic logic has become a familiar face in computer science, not least in the AI and agents-community. However, epistemic logic to reason about communication protocols, especially protocols where matters of security and authentication play an important role, is still mostly unknown territory. If one tries to model a simple security protocol, there are often many aspects of such a protocol that are hard to model or formalize. Assumptions about who plays the role of intruder for instance easily end up being common knowledge in the model, and notions like freshness of a key often receive an ad hoc interpretation. It seems to me epistemic logic for security protocols is still waiting for a break-through.

I have recently seen promising work in Synthese that gave a possible world semantics to a general theory of epistemology, and other work that combines learnability theory and epistemic logic. I felt both examples were harvesting fruit by academics who made an effort to connect different disciplines. Another example where I expect interesting results when researchers with different backgrounds get more interested in each other and learn to collaborate, is in the area of cognitive psychology and epistemic logic.

15

Wolfgang Lenzen

Professor
University of Osnabrück, Germany

---

1. Why were you initially drawn to epistemic logic?

My interest in epistemic logic was raised in the early 70ies by a lucky coincidence. Having completed my doctoral dissertation on a topic in the philosophy of science but still being a "greenhorn" in almost every other field of philosophy, I somehow hit upon Hintikka's book "Knowledge and Belief". I immediately became fascinated by his pioneering attempt to clarify important epistemological issues by way of a thorough logical analysis. As a student, I had attended some courses in classical epistemology, e.g. on Plato, Kant and Hume, and I had tried to understand what were the real issues, e.g. concerning the nature of knowledge, the "problem" of skepticism, or the impossibility to "prove" the existence of an outer world. But, frankly speaking, I never got the feeling of really understanding what it was all about. After all, I had received my education mainly in mathematics, not in philosophy, and it was only due to a lucky coincidence that I eventually gave up the idea of becoming a school teacher for mathematics and decided instead to try the risky career of a professor of philosophy.

During the winter term 1968/69, when I was still a student of mathematics and philosophy at the University of Munich, I learned to my surprise that there exist strong and important connections or "bridges" between these two disciplines. Lectures and seminars by Wolfgang Stegmueller and by Franz von Kutschera showed me that it is possible, or even necessary, to discuss philosophical problems from a rigorous logical point of view. Suddenly I became able, and felt challenged, to participate actively in philosophical discussions, while during the first years of my academic education I had always remained a silent observer, or perhaps admirer, of those sophisticated, but hard to follow, debates about Hegel, Kant, Sartre, and other "deep" philosophers.

As chance would have it, von Kutschera had just received a call for a professorship at the University of Regensburg, and he was looking for candidates for the job of a student assistant and/or research assistant. One Thursday evening, after my presentation of a seminar paper on Goodman's "Structure of Appearance", he asked me if I was interested in taking one of these jobs. During a period of two or three years, this would give me the opportunity to write a doctoral dissertation. It didn't take me a long time to decide to accept this surprising offer, and so I became a philosopher.

In the 70ies the focus of German Analytic Philosophy was almost exclusively centered on logic, philosophy of science, and philosophy of language. Especially for a student of Stegmueller and von Kutschera it had been quite natural to write his doctoral dissertation in philosophy of science. But after I had successfully taken the first step of the academic career by completing a study on "Theorien der Bestätigung wissenschaftlicher Hypothesen", I lost my interest in philosophy of science and looked for something different. Now, while in the Anglo-Saxon community a lively discussion of formal epistemological problems was going on since around 1960, in Germany only very few analytically oriented philosophers would be dealing with such issues. Therefore it was really a lucky coincidence that I hit upon Hintikka's pioneering work and decided to take epistemic logic as subject matter for my Habilitationsschrift.

A first small result of my research about "Probabilistic interpretations of epistemic concepts" was presented at a 1975 conference on "Logic, Methodology, and Philosophy of Science" in London, Ontario, Canada. In the subsequent years I worked my way through the pertinent literature and published a survey of "Recent Work in Epistemic Logic" which appeared in a 1978 issue of Acta Philosophica Fennica. Finally, the Habilitationsschrift "Glauben, Wissen und Wahrscheinlichkeit" containing a detailed study of various systems of epistemic logic was published in 1980 by Springer Verlag, Vienna. Having done all this, however, I again lost interest in this subject and turned to quite different topics: Bioethics, Leibniz's logic, and Philosophy of mind.

2. What example(s) from your work, or work of others, illustrates the relevance of epistemic logic?

As regards the relevance of my own findings, I would normally say that others should better judge on that issue. Yet let me point out to a recent paper of mine where it was shown that epistemic logic may well help to clarify certain basic issues of (visual) perception. In particular, with the help of some intuitive assumptions concerning the knowledge of one's own mental states (usually referred to as the principle of "privileged access"), the commonsense wisdom "Seeing is believing" – indeed even the strengthened version "Seeing is knowing" – becomes a theorem of a suitably expanded epistemic logic (see my contribution "Wahrnehmung" for P. Kolmer & A. Wildfeuer (eds.), *Neues Handbuch philosophischer Grundbegriffe*, Freiburg, Alber Verlag, 2010).

Also, it had turned out earlier that some of my results in the logic of "weak" belief have interesting applications in the field of value logic. Given a subjective probability function $\text{prob}(a, p)$, "weak" belief can be defined by $B(a, p) := \text{prob}(a, p) > \text{prob}(a, \neg p)$. Hence weakly believing that p – in contrast to being entirely convinced that p – only requires that subject a thinks p to be more probable than not. Similarly, in the field of value logic, one may presuppose the existence of a normalized utility function $\text{util}(a, e)$ which represents the value of an event e for subject a in such a way that $\text{util}(a, e) > 0$ if and only if e is preferred or wanted by a, while $\text{util}(a, e) < 0$ if and only if a prefers e not to happen. Accordingly, a "weak" concept of goodness can be defined as $G(a, e) := \text{util}(a, e) > \text{util}(a, \neg e)$, i.e. an event or outcome e is good for a (in a much weaker sense than being, for instance, optimal) just in case it is better for a if e happens rather than not. Now, although there are big differences between the logics of the two "weak" notions, it turned out, somewhat surprisingly, that the key principle for a complete axiomatization of the logic of $G(a, e)$ can be straightforwardly obtained from a corresponding principle for the logic of $B(a, p)$ – see my paper "On the representation of classificatory value structures" in *Theory and Decision* 15 (1983), 349-369.

3. What is the proper role of epistemic logic in relation to other disciplines, for instance mainstream epistemology, game theory, computer sciences or linguistics?

Having participated in a series of conferences on "Theoretical Aspects of Reasoning about Knowledge" organized by the IBM corporation in California, I have come to realize that there exist interesting applications of epistemic logic for problems of representation of "knowledge" (i.e. data) in computer science, but I am not an expert in these issues. As a philosopher I just want to emphasize that epistemic logic still plays the important role of an "organon", i.e. a formal tool, for the analysis of many different concepts both within "classical" epistemology and within philosophy of mind.

4. Which topics and/or contributions should have had more attention in late 20th century epistemic logic?

It seems to me that the distinction between "weak" and "strong" belief (and also the investigation of further concepts between these border cases) should have deserved more attention. In particular, I miss a closer discussion of the principle $C(a,p) \leftrightarrow B(a,K(a,p))$ stating that some person a is convinced that p if and only if a believes that she knows that p. This principle appears very plausible in German where the phrase "a glaubt zu wissen, dass p" is often used by a speaker b ($\neq a$) to express that, although a is entirely convinced that p, she doesn't know that p since, as b himself believes to know, p is not the case, after all.

It would be worthwhile investigating whether in other languages (English, French, Italian, Spanish, etc.) similar phrases exist which confirm the validity of the crucial principle $C(a,p) \leftrightarrow B(a,K(a,p))$.

5. What are the most important open problems in epistemic logic and what are the prospects for progress?

Let me just reply in a Socratic manner: I am sure there will be some important open problems, but I don't know them.

16
John-Jules Ch. Meyer

Professor

University of Utrecht, The Netherlands

---

1. Why were you initially drawn to epistemic logic?

I remember I first heard about epistemic logic at a conference on theoretical computer science (my PhD research subject at the time) in the mid 80s. There was an extra session on the topic by Halpern & Moses, and Joe Halpern presented their well-known seminal IJCAI paper. Having been interested in modal logic, and especially deontic logic, since my master student days, I was immediately interested. I did not yet know the philosophical work on epistemic logic, but this computer science-oriented treatment of the subject was really attracting my attention. Since I was also interested in specification and verification of computer programs and concurrent programming, epistemic logic, Halpern & Moses style, was really appealing to me. Around that time I also came across Ray Turner's little but inspiring book "Logics for Artificial Intelligence", which made me turn into that direction. Although briefly, also in this book epistemic logic was mentioned and it pointed to Bob Moore's work on a logic of knowledge and action. This made me even more interested, since I had been working on dynamic logic as a basis for deontic logic.

Later, being already a staff member at the Vrije University in Amsterdam myself, I followed a course on logics for AI given by Johan van Benthem at the 'other university', the University of Amsterdam. I then thought that this was really interesting, and we should have a course on this at our university as well. I discussed this with my colleagues there, and we decided to give a course on more advanced logic, featuring logic programming (given by my colleague Jan Willem Klop) and epistemic logic (given by me). We also started to develop course material. Of course, we started with an adaptation and extension of the seminal paper by Halpern

& Moses. Jan Willem, who had produced already many papers himself on a type writer, had a small Mac at that time on which the first versions of the lecture notes were produced, I guess in MacWrite, later to be replaced by the first versions of Microsoft Word. Later I got a Mac myself and the material on epistemic logic grew into the book that my former PhD student Wiebe van der Hoek and I published with Cambridge University Press in 1995. The arrival of Wiebe, by the way, was the reason that I got interested in epistemic logic as a research topic rather than just a topic to teach.

Wiebe did his thesis on epistemic logic covering technical issues such as completeness of logics of more advanced epistemic operators such as common knowledge and implicit (distributed) knowledge, as well as operators for graded forms of knowledge. Wiebe and I would work together for some 14 years on the topic of epistemic logic and related topics within the area of logics for AI, such as non-monotonic reasoning, first in Amsterdam, later in Utrecht.

We also got to work with our first joint PhD student Bernd van Linder, with whom we did our first work on intelligent agents: the KARO framework. This was a blend of epistemic and dynamic logic, inspired by Moore's work, and extended into the direction of abilities, goals and commitments of agents, which later in retrospect could be viewed as an alternative BDI (belief-desire-intention) logic based on dynamic logic. This work in turn has inspired us to develop agent-oriented programming languages such as 3APL and 2APL. Even now epistemic operators still play an important role in our work (done jointly with my former PhD students Koen Hindriks and Birna van Riemsdijk, and with Mehdi Dastani, Natasha Alechina and Brian Logan) on modal logics (or at least modally inspired logics) for the verification of agent programs.

2. What example(s) from your work, or work of others, illustrates the relevance of epistemic logic?

As explained above I first encountered epistemic logic through Halpern & Moses' seminal paper. This presented a general introduction to the topic. In subsequent papers, and notably their book, Halpern et al. developed a true computer science perspective on epistemic logic via the so-called interpreted system approach. This conveyed the idea of considering a distributed computer system with several processors connected with each other by a network from an epistemic perspective: what do the processes 'know' about each other? This question is crucial for the analysis of the behaviors of a large class of information processing protocols in such systems. Interpreted systems later also got their way to deontic logics through the work of Lomuscio & Sergot.

We have ourselves also been looking at other applications in CS and AI systems. E.g. for specifying and reasoning about concurrent programs (with my student Marten van Hulst), default / nonmonotonic reasoning systems (e.g. via the study of autoepistemic logic AEL but also - in cooperation with Jan Treur - by using temporal logic to render the nonmonotonic reasoning process 'monotonic'), analogous reasoning (with my master's student Caroline van Leeuwen), reflective architectures (with Lluis Godo, Carles Sierra and Jan Treur), and later agent systems in general (with Bernd van Linder), and agent programs in particular (with Koen Hindriks, Birna van Riemsdijk, Mehdi Dastani, Natasha Alechina and Brian Logan). We also used epistemic logic for analyzing counterfactual reasoning, for a semantic analysis of the linguistic conjunction 'but', and recently even for a conceptual analysis of emotions (the latter with my Ph.D. student Bas Steunebrink).

3. What is the proper role of epistemic logic in relation to other disciplines, for instance mainstream epistemology, game theory, computer sciences or linguistics?

I believe epistemic logic plays an important role in all of these. Being a computer scientists, I've no experience myself in epistemology research. I myself have been mainly involved in epistemic logic for computer science and AI. As stated before, there is a strong relation between computer science / artificial intelligence and epistemic logic, starting with the work on interpreted systems by Halpern et al., but extended to other subfields as well. As mentioned above we ourselves have been involved in some of this as well. One now also sees all sorts of work emerging relating to game theory where a definite epistemic dimension is present (e.g. the work done in Amsterdam by people around Van Benthem and in Liverpool by Van der Hoek and Wooldridge on ATEL). This appears to be an active subfield that will go on for some time... I'm not aware of work done in linguistics where epistemic logic is used other than our own little excursion to the 'logic of but'.

4. Which topics and/or contributions should have had more attention in late 20th century epistemic logic?

This is a hard question, almost a kind of 'what if' question: what if a certain topic would have been picked up? But what comes to mind when thinking about this, is perhaps the more pragmatic side of the use of epistemic logic in computer science, including AI. What I mean here is that normally epistemic models are rather 'ideal', in the sense that agents have infinite resources and are perfect reasoners with these. Of course, there has been work on limited and resource-bounded forms of epistemic reasoning (e.g. work by Fagin & Halpern, Konolige and Alechina & Logan), but I think this topic still has remained a bit marginal seen from the perspective of epistemic logic / reasoning as a whole, while I believe it is a truly important issue if one really wants to employ epistemic logic for reasoning about 'real' agents (and agent programs).

5. What are the most important open problems in epistemic logic and what are the prospects for progress?

As I see it, besides the issue of resource-boundedness, as stated above, the magic word in epistemic logic is now *'dynamics'*. Dynamic epistemic logic is intended to reason about the dynamics of knowledge, i.e. how knowledge changes in multi-agent systems. As multi-agent systems are becoming increasingly important in both academic research and in practice, and for some of these systems it is imperative that we can reason about their behavior to show that they behave correctly / adequately, logics for MAS will also become increasingly important, and these will certainly contain an epistemic component.

(Interesting here is also the view of epistemics being just one dimension next to issues such as motivations (goals) and other mental attitudes, which are also dynamic! So I believe that we should look at systems beyond mere epistemic logic.) As I see it, these logics are not entirely crystallized, also not conceptually, so there is still a lot of interesting work to be done here.

References

N. Alechina, M. Dastani, B. Logan & J.-J. Ch. Meyer, A Logic of Agent Programs, in: Proc. AAAI-07 (R.C. Holte & A.E. Howe, eds.), Vancouver, Canada, AAAI Press, 2007, pp. 795-800.

N. Alechina & B. Logan, Ascribing Beliefs to Resource-Bounded Agents, Proc. of the First International Joint Conference on Autonomous Agents and Multi-Agent Systems (AAMAS 2002), 2002, pp. 881-888.

M. Dastani, 2APL: A Practical Agent Programming Language, Autonomous Agents and Multi-Agent Systems 16(3), 2008, pp. 214-248.

M. Dastani, B. van Riemsdijk & J.-J. Ch. Meyer, A Grounded Specification Language for Agent Programs, in Proc. 6th Int. J. Conf. On Autonomous Agents and Multi-Agent Systems (AAMAS'07) (M. Huhns, O. Shehory, E.H. Durfee & M. Yokoo, eds.), Honolulu, Hawai'i, USA, 2007, pp. 578-585.

R. Fagin & J.Y. Halpern, Belief, Awareness, and Limited Reasoning, in Proc. IJCAI-85, Los Angeles, 1985, 491-501.

L. Godo, W. van der Hoek, J.-J. Ch. Meyer & C. Sierra, Many-Valued Epistemic States: An Application to a Reflective Architecture: MILORD-II, in: Advances in Intelligent Computing - IPMU'94 (5th Int. Conf. on Information Processing and Management of Uncertainty in Knowledge-Based Systems, Paris, July'94 (Selected papers)) (B. Bouchon-Meunier, R.R. Yager & L.A. Zadeh, eds.), LNCS 945, Springer Verlag, Berlin, 1995, pp. 440-452.

J.Y. Halpern & Y. Moses, Guide to the Modal Logics of Knowledge and Belief, in: Proceedings of the 9th International Joint Conference on Artficial Intelligence (IJCAI-85), 1985.

K.V. Hindriks, F.S. de Boer, W. van der Hoek & J.-J. Ch. Meyer, Agent Programming in 3APL, in Int. J. of Autonomous Agents and Multi-Agent Systems 2(4), 1999, pp. 357-401.

K.V. Hindriks & J.-J. Ch. Meyer, Agent Logics as Program Logics: Grounding KARO, in 29th Annual German Conference on AI, KI 2006 (C. Freksa, M. Kohlhase, K. Schill, eds.)), Bremen, Germany, June 14-17, 2006, Proceedings, LNAI 4314, Springer, 2007, pp. 404-418.

K.V. Hindriks & J.-J. Ch. Meyer, An Agent Program Logic with Declarative Goals, in Proc. FAMAS06 (ECAI2006 Workshop on Formal Aspects of Multi-Agent Systems (B. Dunin-Keplicz & R. Verbrugge, eds.), ECCAI, 2006, pp. 1-15.

K. Konolige, A Deduction Model of Belief, Pitman / Morgan Kaufmann, London / Los altos, 1986.

A. Lomuscio & M. Sergot, Deontic Interpreted Systems, Studia Logica Vol. 75 (Special Issue on The Dynamics of Knowledge) (W. van der Hoek & M. Wooldridge, eds.), 2003.

J.-J. Ch. Meyer, Epistemic Logic, Chapter 9 of: The Blackwell Guide to Philosophical Logic (L. Goble, ed.), Blackwell Publishers, Oxford, UK, 2001, pp. 183-202.

J.-J. Ch. Meyer, Modal Epistemic and Doxastic Logic, in: Handbook of Philosophical Logic (2nd edition) (D. Gabbay & F. Guenthner, eds.) Vol. 10, Kluwer, Dordrecht, 2003, pp. 1-38.

J.-J.Ch. Meyer & W. van der Hoek, Non-Monotonic Reasoning by Monotonic Means, extended abstract in J. van Eijck (ed.), Logics in AI (Proc. JELIA '90), LNCS 478, Springer, 1991, pp. 399-411; full version under the title "A Modal Logic for Nonmonotonic Reasoning" in: Non-Monotonic Reasoning and Partial Semantics (W. van der Hoek, J.-J.Ch. Meyer, Y.H. Tan & C. Witteveen, eds.), Ellis Horwood, Chichester, 1992, pp. 37-77.

J.-J. Ch. Meyer & W. van der Hoek, Epistemic Logic for AI and Computer Science, Cambridge Tracts in Theoretical Computer Science 41, Cambridge University Press, 1995.

J.-J. Ch. Meyer & W. van der Hoek, A Default Logic Based on Epistemic States, Fundamenta Informaticae 23(1), 1995, pp. 33-65.

J.-J. Ch. Meyer & W. van der Hoek, Counterfactual Reasoning by (Means of) Defaults, in Annals of Mathematics and Artificial Intelligence 9 (III-IV), 1993, pp. 345-360.

J.-J. Ch. Meyer & W. van der Hoek, A Modal Contrastive Logic: The Logic of 'But', Annals of Mathematics & Artificial Intelligence Vol. 17 (3,4), 1996, pp. 291-313.

J.-J. Ch. Meyer, W. van der Hoek & B. van Linder, A Logical Approach to the Dynamics of Commitments, AI Journal 113, 1999, 1-40.

J.-J. Ch. Meyer & J.C. van Leeuwen, Possible World Semantics for Analogous Reasoning, in: Practical Reasoning (Proc. FAPR'96) (D.M. Gabbay & H.J. Ohlbach, eds.), LNAI 1085, Springer, Berlin, 1996, pp. 414-429.

R.C. Moore, A Formal Theory of Knowledge and Action, in: J.R. Hobbs & R.C. Moore (eds.), Formal Theories of he Common Sense World, Ablex Pub. Corp., 1984.

B. Steunebrink, M. Dastani & J.-J. Ch. Meyer, A Logic of Emotions for Intelligent Agents, in Proc. AAAI-07 (R.C. Holte & A.E. Howe, eds.), Vancouver, Canada, AAAI Press, 2007, pp. 142-147.

R. Turner, Logics for Artificial Intelligence, Ellis Horwood / Wiley, Chichester, 1984.

H. van Ditmarsch, W. van der Hoek & B. Kooi, Dynamic Epistemic Logic, Springer, Heidelberg, 2007.

W. van der Hoek, Modalities for Reasoning about Knowledge and Quantities, PhD thesis, 1992.

W. van der Hoek, M. van Hulst & J.-J. Ch. Meyer, Towards an Epistemic Approach to Reasoning about Concurrent Programs, in: Semantics - Foundations and Applications, Proc. REX Workshop Beekbergen 1992 (J.W. de Bakker, W.P. de Roever & G. Rozenberg, eds.), LNCS 666, Springer, Berlijn, 1993, pp. 261-287.

W. van der Hoek, B. van Linder & J.-J. Ch. Meyer, An Integrated Modal Approach to Rational Agents, in: M. Wooldridge & A. Rao (eds.), Foundations of Rational Agency, Applied Logic Series 14, Kluwer, Dordrecht, 1998, pp. 133-168.

W. van der Hoek & J.-J.Ch. Meyer, Possible Logics of Belief, Logique & Analyse 127-128, 1989, pp.177-194.

W. van der Hoek & J.-J. Ch. Meyer, Making Some Issues of Implicit Knowledge Explicit, Int. J. of Foundations of Computer Science 3(2), 1992, pp. 193-223.

W. van der Hoek & J.-J. Ch. Meyer, Graded Modalities in Epistemic Logic, Logique et Analyse 133-134 (Special issue on Int. Symposium on Epistemic Logic), 1991, pp. 251-270.

W. van der Hoek & J.-J. Ch. Meyer, A Complete Epistemic Logic for Multiple Agents: Combining Distributed and Common Knowledge, in: Epistemic Logic and the Theory of Games and Decisions (M.O.L. Bacharach, L.-A. Gérard-Varet, P. Mongin & H.S. Shin, eds.), Kluwer, Dordrecht, 1997, pp. 35-68.

W. van der Hoek, J.-J. Ch. Meyer & J. Treur, Formal Semantics of Temporal Epistemic Reflection, in Logic Program Synthesis and Transformation - Meta-Programming in Logic, 4th Int. Workshops, LOPSTR'94 and META'94, Pisa, 1994, (L. Fribourg & F. Turini, eds.), LNCS 883, 1994, pp, 332-352.

W. van der Hoek, J.-J. Ch. Meyer & J. Treur, Temporalizing Epistemic Default Logic, JOLLI 7(3), 1998, pp. 341-367.

W. van der Hoek, J.-J. Ch. Meyer & J. Treur, Formal Semantics of Meta-Level Architectures: Temporal Epistemic Reflection. International Journal of Intelligent Systems, vol. 18, 2003, pp. 1293-1317.

M. van Hulst & J.-J. Ch. Meyer, An Epistemic Proof System for Parallel Processes, in Proc. TARK V (R. Fagin, ed.), Pacific Grove, Morgan Kaufmann, San Francisco, 1994, pp. 243-254.

17

Lawrence S. Moss

Director, Program in Pure and Applied Logic
Professor, Mathematics
Adjunct Professor: Informatics, Linguistics
and Philosophy
Indiana University, Bloomington, IN, USA

I am pleased to have an opportunity to think about epistemic logic and my involvement with it, and so I'd like to start by thanking Vincent Hendricks and Olivier Roy for giving me this opportunity, and for their patience on it.

Some beginnings come by chance, but my interaction with epistemic logic was over-determined. I began my undergraduate studies at UCLA in 1976, and my first quarter was devoted to beginning Logic taught in the Philosophy Department using Kalish and Montague's textbook *Logic: Techniques of Formal Reasoning*, Linear Algebra, and introductory Linguistics. My undergraduate years were mostly devoted to mathematics, with a special interest in mathematical logic, and also natural language semantics and theoretical computer science. I think what drew me then is the same thing that later drew me to epistemic logic and other areas: the possibility that mathematics might be not only beautiful but also might have something to teach us about the social world.

My undergraduate years shaded into graduate school, also at UCLA, where I largely pursued the same general topics and eventually wrote a dissertation in mathematics on generalized recursion theory close to the border of set theory. I didn't realize at the time that generalized recursion theory would soon be a *passe* topic in mathematical logic, but this doesn't matter at all. Looking back, what I learned most from my advisor, Yiannis Moschovakis, was a set of skills that were applicable to research on pretty much any topic. At the time, his interest was shifting from descriptive set theory to more foundational matters concerning algorithms, and he also was taking part in discussions with his old friend Jon Barwise at Stanford. Later he would turn in earnest to topics in natural language semantics and in philosophy of language. His connection to Barwise was almost certainly a big help in my spending a year at CSLI. It was there that I first learned about modal logic in the first place: I simply had not had classes in the subject and hardly knew anything about it. That year I also took part in Peter Aczel's seminar on Non-Well-Founded Sets; this directly lead to Barwise and Etchemendy's book *The Liar*, to Jon's work on common knowledge and circularity, to my own involvement with coalgebra, and so on. I also interacted that year with Joseph Goguen and José Meseguer on abstract data type computation theory and other things, and I was generally exposed to issues and ideas that are still very much of live interest. Closer to our topic of epistemic logic, I believe that this was during the same period and same part of the world that Fagin, Halpern, Moses, and Vardi were working on *Reasoning About Knowledge.*

I hopped around from post-doc to post-doc, and from topic to topic. A few years later I got to know Rohit Parikh. I don't remember exactly how we started working on the topic of epistemic logic, but we did. Rohit's intuition was that branches of mathematics really aren't about what they say they are about; the official descriptions sometimes even hide what is really going on. Further, topics in general topology and basic computability theory had a knowledge-theoretic flavor. The idea of our *topologic* was to formalize that "epistemic core" that seemed to lie at the heart of general topology and also computability theory. He also emphasized something to the effect that basic human abilities could (and should) be formalized in logical systems which are decidable and hopefully of low computational complexity. I think of this point as a methodological move made in areas of computer science, cognitive science, and other areas, and I'm generally sympathetic to it. It plays a role in epistemic logic, as I'll mention below. In a different direction, I also advocate studying decidable fragments of natural language. I trace my views on the importance of decidability to Rohit Parikh. He specifically felt that decidability results for topologic were not simply in the category of "Oh, that's nice," but instead that if the logical system for topologic turned out to be undecidable, then the whole thing would be much less interesting.

I visited the CUNY Graduate Center quite often, back in the days when it was on 42nd St. On one of those visits, I remember

Melvin Fitting's student Jan Plaza speaking about his work on a formalization of the epistemic puzzle Mr. Sum and Mr. Product. As far as I know, this was the first time that anyone had attempted to incorporate dynamic operators for public announcements into epistemic logic. Little did I know then that a few years later I would work in the same area. That involvement came about when I heard Jelle Gerbrandy lecture on his work, first at the 1996 edition of the Situation Theory and its Applications conference, and later in the logic seminar at our university. (The STA conferences are no longer running.) I became quite interested, and I must admit that the open completeness problem for the logic of public announcements with common knowledge operators was a strong attraction. I thought it would be fun to work on this with Sławomir Solecki and Alexandru Baltag; Slawek was then at Indiana and the two of us had not worked together, and Alexandru was at the end of his graduate studies. The completeness result turned out to be too easy for Slawek: he probably had no knowledge of modal logic to begin with, but as soon as he saw the Kozen-Parikh argument for the completeness of PDL, he saw that it would work for the logic of public announcements. The more interesting work came after, when we started to think about private announcements. Fairly soon, Alexandru came up with the idea of action models and the action product operation, and these are the centerpieces of dynamic epistemic logic.

So all of this is a long explanation of how I was drawn to epistemic logic. Overall, the draw is both that I like the technical side and I like the fact that it is supposed to be about human knowledge and communication. I am especially fond of dynamic epistemic logic and its many offshoots. I want to shift gears to now talk about these subjects themselves. For me, the main contribution of the subject is via the following set of slogans.

Epistemic Logic Thesis

Let S be a real-world situation involving people and atomic sentences. Then there is a finite set T of natural language sentences which explains S in a precise enough manner so that all possible statements about knowledge, higher-order knowledge, and common knowledge have a determined truth value.

Furthermore, corresponding to S is a mathematical object (specifically, a finite multi-agent Kripke model) $\hat{S}$, a world $s_0 \in \hat{S}$, and a set $\hat{T}$ of sentences in the formal language of multi-agent epistemic logic $\mathcal{L}$ such that for each natural language sentence A about the same atomic facts and the same participants in S, the following are equivalent:

1. A is intuitively true about S.
2. $s_0 \models \hat{A}$ in $\hat{S}$.

Continuing, the set of sentences of $\mathcal{L}$ true at all worlds of all models is *decidable*. Frequently one adds additional requirements to the models to reflect extra intuitions about knowledge or belief, and in those cases one usually again has decidability in this sense.

Finally, corresponding to actions in the real world which change situation to situation, there are finite sets of natural language sentences, and also formal objects which again correspond in much the same ways.

A thesis or an aspiration?

We call this a "thesis" on a parallel with statements like the Church-Turing Thesis in computability. But unlike those, this statement is not discussed in this form in the literature, at least not to my knowledge. The handbook chapter [BDM] comes close. The thesis was certainly in my mind for a number of years. My sense is that part of the reason why it has not been stated is that people in the area simultaneously believe and disbelieve it. Indeed, it could have been stated as the Epistemic Logic Aspiration; that formulation would have been much less controversial. One of the challenging aspects of this interview is exactly this tension between the two aspects. My view is that both formulations are correct – in different ways, of course.

Example

Let us consider the most basic situation: two people and one concealed coin. Agents a and b enter a room. The room has a table some distance from the door, and on the table is a remote-control mechanical coin flipper. Then a presses a button, and the coin spins through the air, landing in a small box on a table. The box closes, and a and b are much too far to see the coin. In reality, the coin shows heads. We are to think of the situation at this point in time.

The agents are a and b, and the only atomic sentence is "the coin shows heads." (We could also take "the coin shows tails" as a different atomic sentence, and this would lead to different details all the way through.) The sentences above are not a situation but rather describe one. So they play the role of (a subset of) T in the statement of the Epistemic Logic Thesis. This is typical, because we are unable with printed words to present S. In addition to the sentences above, we also need some others in order to satisfy the requirement that T be detailed enough to intuitively entail all statements about knowledge, including higher-order knowledge and common knowledge. Here is what we take: It is common knowledge to A and B that neither knows the state of the coin.

The mathematical model $\hat{S}$ is the model shown below:

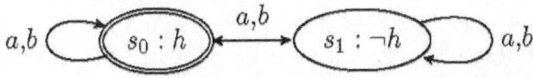

The world s_0 is the designated one. An example of a sentence A in English would be "a knows that b doesn't know that the coin shows heads." We would translate this as $\hat{A}$, the formal sentence $K_a \neg K_b h$. Then A is intuitively true about the described situation s, and $\hat{A}$ is formally true of the world s_0 in the model $\hat{S}$. Of course, there are other examples of sentences in English and their translations, and what the Thesis asserts is that intuitive truth of A is equivalent to formal truth of the translations.

Discussion

As I mentioned, the terminology of a "thesis" makes a parallel with the Church-Turing Thesis in computability theory. The thesis here asserts that to some extent our natural language descriptions of situations involving knowledge are adequately formalized using the semantic models and formal languages of epistemic logic. This thesis is certainly open to debate, and I shall shortly rehearse a version of the arguments against it. In addition, almost as soon as one writes the thesis itself down, one is tempted to make changes and additions, or to make exceptions. (So here the parallel to computability breaks down.) But the value of epistemic logic for me is precisely that involves enough material to state this kind of thesis and argue for it, and at the same time it leads to some additional work involving refinements of the thesis, and to opposing points.

Usually one does not begin with a real-world situation on its own in the first place but rather with the finite set T of natural language sentences, and then S would be an abstract object living intersubjectively. One could re-state the thesis in that form, and then the requirement on T becomes an *assumption*. This assumption on T would then state that T is detailed and precise enough to lead someone who heard T would have a definite intuition about all statements involving knowledge and the basic sentences. In the real world, descriptions are almost never this complete. One then could wonder about whether the thesis itself ever applies. But this point applies more generally, and it merely causes us to gape at the gap between the world and our formalizations of it.

A different point: the way I have stated the thesis allows for statements of common knowledge to be part of the set T. Indeed, as the example suggests, it is *necessary* to use statements of common knowledge in order to explain S in the sense of the thesis. This is a technical point that I do not wish to explain here. And a final point related to this: if we start not with S but with T, and if T is rich enough, then it will have a model which is unique up to bisimulation.

Logical Omniscience

One way to dispute the thesis comes reflects other issues having to do with the translation between natural language and logic. I have in mind the discrepancy between normative uses of the conditional and the formalization using the material conditional. This suggests that the usual translation $A \mapsto A^*$ of natural language into logic doesn't work the way it should, even for the case of sentences which do not mention knowledge.

But because this objection is made on grounds that are not special to epistemic logic, I will set it aside. Instead, let us focus on a different objection that raises live issues for the subject.

No discussion of epistemic logic can ignore the philosophical problems that are at the heart of it. One major problem with the Epistemic Logic Thesis is that there is no explanation of *how* the mathematical model $\hat{S}$ comes to represent S. But this is a problem with mathematical models in general. Perhaps a more salient set of objections to the thesis comes from the topic of *logical omniscience*. Let T be a finite set of sentences in natural language describing a situation S. The thesis gives us a models $\hat{S}$ and a world s_0 in it, and in s_0 the translations of sentences of the form "a knows A" are always true provided that A is a tautology of propositional logic. This flies in the face of our insistence that T describes a real situation, where agents certainly do not know all tautologies. What are we to make of this discrepancy? One way out is to call on something like the *competence/performance* distinction in linguistics: it is a matter of competence that people *do* know all tautologies, but in practice, they do not. But I am not satisfied with this kind of move, and I am sure that others also find it unsatisfactory. The problem is that we began with human phenomena, and so we should strive to understand them, no matter how hard this turns out to be.

Interactions with cognitive science

I would like to close this short interview with a suggestion for the future.

I teach a course called Mathematics and Logic for Cognitive Science to graduate students in Cognitive Psychology, Cognitive Science, Artificial Intelligence, and other areas. Most of the course is on topics related to the mathematical models that most of my students will see, such as Hopfield Nets, Bayesian Nets, Hidden Markov Models, the Singular Value Decomposition of Matrices, Entropy, and Markov Decision Processes. I need to mention that some of the students are downright hostile to the ideas of representation in the first place, and others reject the syntax/semantics distinction. Some also have heard of the Gödel Incompleteness Theorem and believe that it consigns logic to the dustbin of academe. The logic topics at the end of the semester are my way to advertise the attractive parts of logic to students who are not disposed to like it; and even the others mostly will go on to other topics. So I feel challenged to select the "right" ones. I usually aim for dynamic epistemic logic, and if there's time I try to look at logics for conditionals, and to the doxastic extensions of DEL. I think students really love the topic; in fact, my experience teach modal logic in general is that DEL is a wonderful overall motivation. The fact that the representations mostly *do* work is a big point in its favor: can one seriously try to account for people's intuitions using the kinds of mathematical models I mentioned above? Another point: the semantics of modal logic and the attractive relational models reinforce each other; one sees some of each and pretty soon the whole thing fits together as a package. The fact that there are conceptual difficulties at the core comes with the territory: witness neural nets.

My suggestion, and indeed my own current research aspiration for epistemic logic is to look seriously at areas of contemporary cognitive science to try to then make connections of interest to both sides. Cognitive science is full of settings where one talks about learning, memory, reasoning under uncertainty, belief revision, jumping to conclusions, and the like. For an epistemic logician, these areas offer the chances to both do interesting logic and at the same time re-work the very foundations of the subject.

17.1 REFERENCES

[BDM] Alexandru Baltag, Hans van Ditmarsch, and Lawrence S. Moss. Epistemic Logic and Information Update, P. Adriaans and J. van Benthem (eds.), *Handbook of the Philosophy of Information*, Elsevier, 2008, 369–463.

18
Rohit Parikh

Distinguished Professor

City University of New York – Brooklyn College and CUNY Graduate Center, USA

---

1. Why were you initially drawn to epistemic logic?

It was essentially an accident – I was asked to referee a paper in the area (I believe Albert Meyer was the editor) and seeing that the author was having problems with his material, I became interested. I was also influenced by the work of Halpern and Moses on the co-ordinated attack problem and the work of Chandy and Misra on knowledge in distributed processes. Yet another fact which entered was that I had already been working in Dynamic Logic, especially Propositional Dynamic Logic (PDL), and the techniques and intuitions of PDL easily transferred to the logic of knowledge. It is easy for instance, to give an interpretation of the logic of S5 knowledge into PDL with converse, and then a decision procedure for the latter can be converted into a decision procedure for the former (See [Pa91]).

2. What example(s) from your work, or work of others, illustrates the relevance of epistemic logic?

Epistemic logic is ubiquitous, so it is hard to say where it is not relevant! Just a couple of examples.

Game theory: The work of Aumann on 'Agreeing to disagree' was path breaking. Aumann considered two agents who had started with the same prior probability distribution but had then received different information. Naturally, they would now have different views. But Aumann showed that could not *agree to disagree*.

If, for instance, they were considering the probability of some event, they could now (after taking in the new information) have different values for these probabilities, but these new values could not be common knowledge. If the values were common knowledge, then they must be the same. The agents could not agree to disagree.

Aumann's result was generalized by Bacharach and Cave. However, the question arose what would happen if the probabilities were not common knowledge. This issue was discussed by Geanakoplos and Polemarchakis who showed that if two agents repeatedly exchanged their values, then they would revise their own views and eventually these values would become the same even though the agents had still not disclosed the private information that they had received.

This work was extended by myself and Krasucki. Krasucki and I showed that if many agents communicated in pairs, conveying their latest values of some random variable (satisfying a convexity condition) then eventually these values would become equal. Suppose for instance that two agents A and B had never spoken to each other, but A had spoken to C and B to D, and C and D had spoken to each other (all repeatedly) then not only C and D, but also A and B would end up having the same value. These results showed that the notion of knowledge is important in how consensus about, say, prices, is arrived at through the flow of information.

The work of Milgrom and Stokey is a dramatic example of the power of knowledge theoretic considerations. Milgrom and Stokey showed that under very general assumptions, two profit seeking agents could not trade since it would be common knowledge that they both expected to profit. Suppose A was selling a stock to B, then the fact that the sale was taking place would reveal that A expected the stock to go down relative to the rest of the market and B expected it to go up, and this fact would be common knowledge! This common knowledge is not in fact possible. We know that sales do take place, but perhaps this phenomenon is due to insufficient thought being given to the issues by the two agents, or perhaps due to the fact that the Aumann model does not work perfectly.

Deontic Logic: Another example, in the context of Deontic logic, is the work done by Eva Cogan, Eric Pacuit and myself on knowledge based obligation. In that work we point out that in order for an agent to have an obligation it is necessary for the agent to know that the circumstances are such that the obligation will 'kick in'. A doctor does not have the obligation to treat a sick man unless she knows that he is sick. Similarly, you cannot be expected to help a drowning person unless you are in the situation where you know that someone is drowning and know, moreover, that among the people who could help her, you are the best equipped.

Thus deontic logic needs to be combined with epistemic logic in order to compute an agent's actual obligations. In this work with Cogan and Pacuit, we have suggested that the case of Kitty Genovese would fall under this rubric. In 1964, Kitty Genovese was raped and murdered within sight of more than 30 neighbors, none of whom called the police. This event stunned the nation. What the three of us (Cogan, Pacuit and I) have suggested is that this failure to help was caused by the fact that none of the neighbors had the unique role of calling the police. Experimental evidence has shown that when a *small* group of people witness such an unfortunate event, then it is much *more* likely that one of them will help.

Topologic: In some work started with Larry Moss, I and other co-workers have investigated the knowledge theoretic basis of topology. An open set is like a recursively enumerable (r.e.) set in the sense that if you are in it, you *can* know it. Every element of an r.e. set is eventually enumerated, and so of any element of the set, eventually you know it is in. Similarly, if a point p is inside an open set X then a sufficiently fine measurement of its position will reveal that p is in X. We do not need to know exactly where p is. If we know a small enough neighborhood N of p, then it will lie entirely in X and we can conclude that no matter where p is inside N, p will be in X. A point on the *boundary* of a closed set, by contrast, would appear to retain the possibility being outside, no matter how fine your measurement of its position.

If we think of X as a unary predicate then the formula $X \to \Diamond KX$ expresses the fact that X is open.

This observation reveals a deep connection between topology and epistemic logic and allows us to define topological notions via their knowledge theoretic properties. This insight leads to a decidable logic, and properties, including completeness theorems – have been proved by Dabrowski, Georgatos, Heinemann and Weiss, apart of course from Moss and Parikh (see [PMS] for a survey).

3. What is the proper role of epistemic logic in relation to other disciplines, for instance mainstream epistemology, game theory, computer sciences or linguistics?

I see the relation with game theory, linguistics and animal psychology to be more crucial than the relation with epistemology. The reason is that epistemology is foundational and thus investigates deeper issues than epistemic logic does. Epistemic logic builds on (or should build on) epistemology rather than the other way around. But with the other, more applied areas, the importance of epistemic logic is clear. I have already mentioned how the notion of common knowledge was brought explicitly into game theory by Aumann. The 'no trade' paradox of Milgrom and Stokey is a direct result of Aumann's work. There is also the more 'sexy' work done recently by Brandenburger and Keisler [PacBK] who manage to embed Russell's paradox into the question of how two agents thinking about each other have certain convoluted states of knowledge. Artemov and Fitting have recently offered knowledge theoretic analyses of the Centipede game.

The connection between epistemic logic and linguistics is already there and likely to become closer as time passes. One way to understand language is to see it as an attempt to increase the listener's knowledge. This picture is of course overly simple, and Grice already noticed the strategizing which takes place in conversation. The notion of implicature is essentially a game theoretic notion. If a motorist says to a pedestrian, "My car is out of gas" and the pedestrian responds, "There is a gas station around the corner", the pedestrian has not *said*, but *implicated* that as far as she knows, the gas station is open. That the gas station is open is not part of the meaning but is implied, and the implication is *cancellable* in that the pedestrian could then go on to say, "But I am afraid it is closed at this time of the day."

Grice assumed a speaker and a listener who are co-operative, but more recent work has also considered partial co-operation. Of course, no information can be passed in a zero sum game as the interests of the parties are directly opposed, but information can be passed in a game where the interests (as for instance those of Obama and Karzai) are partially aligned. Obama cannot believe everything which Karzai says to him since the latter must exaggerate the extent to which Afghan forces are making progress. Similarly, Obama cannot be entirely honest with Karzai, since the possibility of the US simply walking away, which is always there, cannot be mentioned lest he lose all co-operation from Karzai. And yet the two do have interests in common and some information needs to be passed in a way which will be trusted – somewhat.

4. Which topics and/or contributions should have had more attention in late 20th century epistemic logic?

Well, I would say that the problem of logical omniscience should have received more attention. Most logics of knowledge stipulate that agents know all logical truths, and that, moreover, they know all the consequences of what they know. These assumptions are of course not realistic. If this were really so, none of us would need to study mathematics, since we would already know everything we were going to derive. Indeed the whole process of reasoning would become superfluous. When we carry out a derivation, the conclusion is already contained (semantically) in the assumptions we start with. But by logical omniscience, whatever we find out was 'already known'. So why bother?

The problem of logical omniscience was already mentioned by Leonard Savage in his 1954 classic *The Foundations of Statistics* (p. 20). As a justification for the assumption of logical omniscience he says, "logic can be interpreted as a crude, but sometimes handy psychological theory."

Yet, the assumption of logical omniscience makes for very elegant formal systems with nice properties. People were so much in love with the cute formal systems being studied, and their properties like completeness, deicidability, complexity and expressive power, that they did not look too closely at the foundations of epistemic logic. This is changing now, but not fast enough. Some sort of momentum causes people to keep on working with formal systems which they know are not satisfactory.

5. What are the most important open problems in epistemic logic and what are the prospects for progress?

I would say that the most important problem is to develop realistic theories of actual knowledge. We know that human beings are not very good at drawing logical consequences from what they already know. They suffer from the defects pointed out by Kahneman in his Nobel lecture of 2002. Kahneman gives examples where even Princeton students make inferences which are patently false.

This lack of logical omniscience can even be of benefit. For instance, work in cryptography is often founded on the notion of a number n which is a product of two large primes p and q. A message coded on the basis of the formula $n = pq$ can be decoded by someone who knows the factorization, but not by someone who does not. And the fact that p and q are so large makes the problem intractable. Our banking is done on the basis of such codes. And yet of course, if we were logically omniscient, then we could factorize n and hence the code would be useless. Online banking is possible only because of our lack of logical omniscience.

Similarly, the notion of *one way keys* involves a lack of logical omniscience. In this application, Ann can code a message for Bob using Bob's public key f_B. The key f_B is public, but its inverse f_B^{-1} is known only to Bob. Thus only Bob can decode the message. This fact is similarly founded on the lack of logical omniscience. A logically omniscient person who knows a function f_B can also know its inverse. However, going from f_B to f_B^{-1} is computationally hard and cannot be solved using moderate computational resources.

A very important area which I hope will receive further attention is the connection between human and animal cognition. In a famous paper, Premack and Woodruff raised the question of the extent to which chimps and other apes have mental representations of the minds of other apes (and people). Work done by Tomasello and his co-workers indicates that the answer is sort of 'not much, but enough to be interesting'. Experimental research by Verbrugge and her colleagues [Ver] shows that even among humans, the level of understanding of others is 'not very much'.

References

[Art] S. Artemov. Knowledge-based rational decisions. Technical Report TR-2009011, CUNY Ph.D. Program in Computer Science, September 2009.

[Aum] R. Aumann, Agreeing to Disagree, *Annals of Statistics*, **4** (1976), 1236-1239.

[Ba] M. Bacharach, Some Extensions of a Claim of Aumann in an Axiomatic Model of Knowledge, *J. Economic Theory*, **37** (1985), 167-190.

[Ca] J.A.K. Cave, Learning to Agree, *Economics Letters* **12** (1983), 147-152.

[CM] M. Chandy and J. Misra, How Processes Learn, *Proceedings* of *4th ACM Conference on Principles of Distributed Computing* (1985) pp 204-214.

[Fi] Melvin Fitting, Reasoning About Games, Technical Report TR-2010002, CUNY PhD Program in Computer Science, March 2010.

[GP] J. Geanakoplos and H. Polemarchakis, We Can't Disagree Forever, *J. Economic Theory*, **28** (1982), 192-200.

[Grice] Paul Grice, *Studies in the ways of Words*, Harvard U. Press (1989).

[HM] Halpern and Moses, Knowledge and Common Knowledge in a Distributed Environment, *Jour. Assoc. Comp. Mach.* v.37 n.3, p.549-587, July 1990)

[Ka] Daniel Kahneman, Maps of Bounded Rationality, Nobel lecture, 2002. http://nobelprize.org/mediaplayer/index.php?id=531

[LL] Loes Olde Loohuis, Obligations in a responsible world, in *Logic, Rationality and Interaction*, Spinger LNAI 5834, edited by He, Horty and Pacuit, 251-262.

I[MS] Paul Milgrom and Nancy Stokey, Information, trade and common knowledge, *Journal of Economic Theory*, Volume 26, Issue 1, February 1982, Pages 17-27.

[PPG] Eric Pacuit, Rohit Parikh and Eva Cogan, The logic of knowledge based obligation in *Knowledge, Rationality and Action*, (2006). **149** 311-341.

[PP] Eric Pacuit and Rohit Parikh, Social Interaction, Knowledge, and Social Software, in *Interactive Computation: The New Paradigm*, ed. Dina Goldin, Scott Smolka, Peter Wegner, Springer publishers, 2007, 441-462.

[Pa86] R. Parikh, Levels of Knowledge in Distributed Computing, *IEEE Symposium on Logic in Computer Science*, Boston 1986, 322-331.

[PK] Parikh, R., Krasucki, P., 1990. Communication, consensus, and knowledge. *Journal of Economic Theory* 52, 178-189.

[KP] D. Kozen and R. Parikh. An elementary proof of the completeness of PDL. *Theoretical Computer Science*, 14:113–118, 1981.

[Lew] Lewis, D.K. (1969). *Convention: A Philosophical Study*. Blackwell Publisher (2002).

[PacBK] Eric Pacuit, Understanding the Brandenburger-Keisler Paradox, *Studia Logica* 86 (2007), 439-454.

[PR85] R. Parikh and R. Ramanujam, Distributed Processes and the Logic of Knowledge, *Logics of Programs*, Springer Lecture Notes in Computer Science, **193**, (1985), 256-268.

[Pa91] R. Parikh, "Monotonic and Non-monotonic Logics of Knowledge", in *Fundamenta Informatica* special issue, *Logics for Artificial Intelligence* vol XV (1991) pp. 255-274.

[PR03] R. Parikh and R. Ramanujam, A Knowledge based Semantics of Messages, *J. Logic, Language and Information* **12** 2003, 453-467.

[PMS] Rohit Parikh, Larry Moss and Chris Steinsvold, Topology and Epistemic Logic, in *Logic of Space*, edited by Johan van Benthem et al, 2007.

[Pa03] R. Parikh, Levels of Knowledge, Games, and Group Action, *Research in Economics* **57** 2003, 267-281.

[PW] D Premack, G Woodruff, Does the chimpanzee have a theory of mind? *Behavioral and Brain sciences*, 1978, 515-526.

[Sav] Leonard Savage, *The Foundations of Statistics*, Dover 1972 (originally published in 1954 by Jophn Wiley and sons).

[Tom] Tomasello, M. (2008). *Origins of Human Communication*, MIT Press.

[Ver] Rineke Verbrugge (2009). Logic and Social Cognition, *Journal of Philosophical Logic* 38 (6).

[WP] Wimmer, H. & Perner, J. (1983). Beliefs about beliefs: Representation and constraining function of wrong beliefs in young children's understanding of deception. *Cognition*, 13, 41-68.

19
Wlodek Rabinowicz

Professor of Practical Philosophy

Lund University, Sweden

---

1. Why were you initially drawn to epistemic logic?

In 1969, after having been expelled from Warsaw university in the aftermath of the student protests in March 1968, I moved to Uppsala, to continue my undergraduate studies in philosophy. I didn't know what I was getting myself into. The department in Uppsala was then one of the main centres for research in modal logic, with people as Stig Kanger, Sven Danielsson and Lennart Aqvist as the leading figures. It was in the air that formalization and axiomatization were the right way to proceed in conceptual analysis. Especially, possible-world semantics was viewed as a panacea for practically all classical philosophical problems. Its spectacular successes in metaphysics suggested it would be equally helpful in dealing with other subjects as well. Deontic logic, epistemic logic, logic of action, etc. all made their appearance on the philosophical scene at that time. Saul Kripke and David Lewis were the new heroes to whom everyone looked up as sources of inspiration.

In Poland I had been strongly influenced by Husserlian phenomenology but I realized I couldn't continue on these lines in Uppsala where phenomenology was viewed with much suspicion. Originally, therefore, my plan was to write a doctoral dissertation on the neo-Kantian philosophy of Ernst Cassirer, but the influence of the new intellectually vibrant environment in which I found myself upon my arrival to Uppsala was too strong to resist. I changed the subject of my thesis several times during these formative years in the 70-ies and each change was traumatic: I felt I was losing time.

One of these failed doctoral projects dealt with logic of acceptance. While working on that subject I wrote a paper on Isaac Levi's *Gambling with Truth*. It was an attempt to account for and expand on Levi's theory using the vocabulary of modal logic. (Rabinowicz 1979) It the end, however, I wrote my dissertation on universalizability in ethics, making heavy use of possible-worlds models. Too heavy: There weren't many moral philosophers who found reading the book worth the effort and, in hindsight, I think they were right. The formal machinery in the thesis was way too sophisticated for the subject I dealt with.

Anyway, after the completion of the thesis I moved to other areas. I made some use of epistemic possible-worlds models in my work on the intuitionistic concept of truth as the possibility of verification (Rabinowicz 1985), but this was just something I did on the side. Influenced by my supervisor, Lars Bergström, I was interested in consequentialist approaches to morality. This rather naturally led me to causal decision theory, which was all the rage in the early 80s. With such topics it was inevitable to go into subjective probabilities. Distinctions between probabilities of subjunctive conditionals, conditional probabilities and the probabilities one would acquire upon learning that the condition obtains all play important role for a causal decision theorist. Similar distinctions can be made for beliefs, which I think was one of the reasons I found theory of belief revision, put forward by Alchourrón, Makinson and Gärdenfors (the famous AGM), such a fascinating area. Another important factor was my friendship with Sten Lindström, who was in Uppsala at that time. Somehow we found we were on the same intellectual wavelength. Sten and I started our long and intensive collaboration on AGM-related issues in the second half of the 80s and from then on I concentrated on the formal approaches to epistemic problems for more than a decade, until the end of the 90s. (See Lindström & Rabinowicz 1989, 1991, 1995, Rabinowicz & Lindström 1994, Rabinowicz 1996.) The arrival of Krister Segerberg, who took up the chair in theoretical philosophy after Stig Kanger, contributed to keeping me on this track, even after I left Uppsala and moved to Lund, where I was appointed to the chair in practical philosophy in 1995. Krister and I did some work together on Dorothy Edgington's solution of Fitch's knowability paradox, using in our approach a two-dimensional semantics for epistemic operators. (Rabinowicz & Segerberg 1994) But Krister's main project in those years was to use AGM as a point of departure for the construction of dynamic doxastic logic (DDL), with a dyadic sentential belief-revision operator interacting with the standard belief operator on sentences. This proved to lead to new interesting problems that kept Sten and me occupied for some time. We were interested in extending DDL to introspective agents who come to acquire information not just about the external world but also about their own doxastic state. (Lindström & Rabinowicz 1999a, 1999b) At about the same time I also got interested in the epistemic aspects of game theory, especially in connection with the use of backward induction as a game-solving method, which I find to be an exciting but difficult area. (Rabinowicz 1998, Broome & Rabinowicz 1999, Jiborn & Rabinowicz 2003.)

In recent years, together with another good friend of mine, Luc Bovens, I did some work on the truth-tracking capacities of judgment aggregation (Bovens & Rabinowicz 2004, 2006). We compared in that regard the premise-based procedure, in which aggregation of individual judgments is applied already at the level of premises of an argument, with the conclusion-based procedure, in which the aggregation step only occurs after each individual has formed a judgment concerning the argument's conclusion. However, my main field of research in those years has been in other areas. Nowadays, I mostly work on formal axiology and value analysis, with occasional forays into decision theory.

2. What example(s) from your work, or work of others, illustrates the relevance of epistemic logic?

Arguing for the relevance of a certain discipline, especially if that discipline is so abstruse as epistemic logic, is difficult. From my own perspective, as a philosopher, a discipline is relevant if it provides useful philosophical insights. I will therefore use as the example one of the results Sten Lindström and I reached together. I think it has this feature of being philosophically illuminating.

In AGM, belief revision is thought of as a function: Given the anterior doxastic state of the agent and a proposition that represents new information, the belief revision operator takes the agent to a new doxastic state. If one thinks of belief change in normative terms, this functional character of AGM revision is problematic. In the interesting cases, ones that AGM focuses on, the new information is incompatible with what the agent originally believes. Incorporating such information isn't therefore straightforward; we cannot simply add it to the stock of our original beliefs and then close the set under logical consequence. So why should it be fully determined what the agent's new belief state is to look like?

The obvious alternative is to go relational, i.e. to allow for the existence of several different doxastic states each of which could be an outcome of the revision of the original doxastic state with a given proposition. The idea is that different revision outcomes might be equally permissible.

Going relational requires weakening of the AGM axioms, but how should we model this new approach? As is well-known, Adam Grove provided a modelling of the functional AGM in terms of a system of spheres that zeroes on the original belief state. The original state K is represented as a set X_K of possible worlds that are compatible with beliefs in K and each sphere in the Groveian model is a superset of X_K – a "fallback", so to speak, to which we might need to retreat in order to accommodate new information, if that information turns out to be incompatible with our original beliefs. X_K itself is the smallest sphere in this system and the whole system is concentric: For any two spheres, one of them is included in the other. If K is revised with a proposition A, the outcome of this revision is taken to be the intersection of A (here understood as the set of possible worlds in which this proposition is true) with the smallest sphere in the system that is compatible with A (i.e. the smallest sphere that intersects this proposition; the case when A is the empty set is treated separately). We retreat in this process to the *smallest* sphere compatible with A in order to make the revision as minimal as possible.

Now, in Lindström & Rabinowicz (1991) we suggest that going relational in belief revision involves giving up the assumption of concentricity for a sphere system. While X_K still is the smallest sphere in our modelling, it is no longer required that for any two spheres in the system one of them must be included in the other. We call such a system a family of fallbacks. In a fallback family, there might be *several* smallest fallbacks that are compatible with the new information A. (They are smallest in the sense that they do not include other fallbacks compatible with A.) For each of these fallbacks, its intersection with A represents a permissible outcome of the revision of K with A.

As is well-known, the Groveian model is equivalent to Gärdenfors' modelling of belief revision in terms of an *entrenchment* ordering of propositions. Intuitively, when we need to revise K with A and for this reason need to remove some propositions from K to make room for A, we keep the more entrenched propositions and let go of the ones that are less entrenched. The propositions that do not belong to K (i.e. the propositions in which X_K is not included) are all minimally entrenched, while logical truths are maximally entrenched. In addition, the ordering of entrenchment is taken to be transitive and complete. The requirement of completeness means that for any two propositions, one of them is at least entrenched as the other, relative to K. In other words, all propositions are mutually comparable as regards their degree of entrenchment. It can be shown that such an entrenchment ordering is definable in terms of the Groveian system of spheres: A proposition B is at least as entrenched as a proposition C iff every sphere included in C is included in B as well. Conversely, the original system of spheres is then recoverable from the so-defined entrenchment ordering: Spheres are sets of possible worlds that are closed upward under entrenchment.

Can this equivalence and mutual definability between entrenchment and the fallback system be retained when we no longer require the latter to be concentric? The answer is yes, provided that we give up the assumption of completeness for the entrenchment ordering. In other words, we have reasons to go relational in belief revision if we no longer suppose that all propositions must be mutually comparable as regards their degree of entrenchment. Since comparability gaps of this kind are only to be expected, on independent grounds, the relational approach appears to be philosophically well-motivated.

Well-motivated, but not mandatory. It is possible to retain the functional approach to belief revision if one opts for an extremely cautious interpretation of this operation. On this cautious account, the revision of K with A only retains what is common to all the revisions of K with A that are taken to be permissible on our relational account. (Formally speaking, this would mean that we take the revision of K with A to be the set of all A-worlds that belong to the union of smallest fallbacks that are compatible with A.) The disadvantage is that such a cautious revision no longer can be said to be minimal: In retaining only the propositions that are common to different plausible candidates for revision, we are bound to give up much of what we have originally believed.

A different way in which the relational approach could be modelled was suggested by Hans Rott (personal communication). On that proposal, we assume that a given K is associated with a *set* of complete entrenchment orderings (or, equivalently, with a set of Groveian concentric sphere systems), rather than with a single ordering containing gaps. For any given A, each of the elements of this set of complete orderings determines a unique revision of K with A. All these revisions are then taken to be equally permissible. (If our gappy ordering is interpreted as the intersection of this set of complete entrenchment orderings, then permissible revisions of K with A on Rott's proposal will include all the revisions that are permissible on our proposal, but not necessarily vice versa.) This kind of modelling is more powerful than the one we have proposed, but it is unclear whether the added power can be put to some philosophical use.

3. What is the proper role of epistemic logic in relation to other disciplines, for instance mainstream epistemology, game theory, computer sciences or linguistics?

Epistemic logic – like, say, classical logic, or alethic modal logic – is of interest in its own right, quite independently of the use it might have for other disciplines. It gives us a deeper understanding of the structural features of our fundamental epistemic concepts. But then this logic might of course also be very useful in various ways for, say, game theorists, computer scientists, political scientists, sociologists, etc. Thus, to take just one example, it can be applied in game theory in the description of the epistemic assumptions that are being made. Not least, theories of belief revision can be applied to account for changes of beliefs about future play that might occur when one is confronted with unexpected moves made by other players (or, for that matter, with unexpected moves made by oneself).

For my own part, though, I am a bit wary of using logical languages in game theory. It seems to me that the logician's *models*, which he comes up with in order to provide the semantics for his logical systems, are more useful for game theorists than the axioms and rules of inference. It is easier to understand what is going on in an interaction between agents if one thinks about such matters in terms of models rather than in terms of principles formulated in some artificial language. Both this language and the models are anyway the theoretician's tools to begin with, rather than the tools of the agents about which we theorize.

The case of computer science is different, though. There the agents themselves are artificial units that can be assumed to reason using formal language.

But perhaps I am too pessimistic. Even when the interacting agents are real people, it might perhaps be illuminating to try to reconstruct their reasoning in an appropriate formal language with epistemic operators.

4. Which topics and/or contributions should have had more attention in late 20th century epistemic logic?

It is a difficult question, but of course one can now partly answer it with the benefit of hindsight. The exciting work on judgment aggregation that was done in the first decade of the 21st century (with Christian List and Franz Dietrich as leading figures in this movement) could probably have been done much earlier. Parallels between this area and the study of collective preferences initiated by Arrow are quite striking and should have been spotted right away. Something similar applies to the debates concerning backward induction. The paradoxical nature of this game-solving method should have been recognized at a much earlier stage: Parallels between backward induction and the student's reasoning in the unexpected examination paradox could have been detected immediately. But it took about forty years, until the end of the 80s, for the lessons from the latter area to be applied in game-theoretical contexts.

5. What are the most important open problems in epistemic logic and what are the prospects for progress?

Again, a difficult question that I am not competent to answer. But, let me just point to one problem, within the broad area of formal approaches to belief change and probability kinematics, that I find quite difficult to handle: What are the principles governing change in our beliefs and in our probabilities, when the new information we acquire is *indexical* in nature? As is well known, updates with indexical propositions lead to special problems. Sleeping Beauty is by now a famous example of this kind. As has been argued by Adam Elga, cases like this can violate fundamental principles of probability kinematics such as the Principle of Reflection. What we would need is a general theory for updates with indexical information, but "to my knowledge" such theory is not yet in sight.

References

Bovens, L. and Rabinowicz, W. (2004) "Voting Procedures for Complex Collective Decisions" An Epistemic Perspective", *Ratio Juris* 17, 241-58.

Bovens, L. and Rabinowicz, W. (2006), "Democratic Answers to Complex Questions: An Epistemic Perspective", *Synthese* 150, 131-53.

Broome, J. and Rabinowicz, W. (1999), "Backwards Induction in the Centipede Game", with John Broome, *Analysis* 59, 237-242.

Jiborn, M. and rabinowicz, W. (2003), "Re-considering the Foole's Rejoinder: Backward Induction in Indefinitely Iterated Prisoner's Dilemmas", *Synthese* 136, 135-57.

Lindström, S. and Rabinowicz, W. (1989), "On probabilistic representation of non-probabilistic belief revision", *Journal of Philosophical Logic* 18, 69-101.

Lindström, S. and Rabinowicz, W. (1991), "Epistemic entrenchment with incomparabilities and relational belief revision", in A. Fuhrmann och M. Morreau (eds.), *The Logic of Theory Change*, Berlin: Springer Verlag, 93-126.

Lindström, S. and Rabinowicz, W. (1995), "The Ramsey Test Revisited", in Gabriela Crocco, Luis Farinas del Cerro and Andreas Herzig (eds.), *Conditionals in AI*, Oxford University Press, 1995, 147-192.

Lindström, S. and Rabinowicz, W. (1998), "Conditionals and the Ramsey Test", in Dov Gabbay and Philippe Smets (eds.), *Handbook of Defeasible Reasoning and Uncertainty Mangement Systems*, vol. 3 (*Belief Change*), Dordrecht: Kluwer, 147-188.

Lindström, S. and Rabinowicz, W. (1999a), "DDL Unlimited - Dynamic Doxastic Logic for Introspective Agents", *Erkenntnis* 50, 1999, 353-385.

Lindström, S. and Rabinowicz, W. (1999b), "Belief Change for Introspective Agents", in Bengt Hansson, Sören Halldén, Wlodek Rabinowicz and Nils-Eric-Sahlin (eds.) *Spinning Ideas: An Electronic Festschrift for Peter Gärdenfors*, http://www.lucs.lu.se/spinning/

Rabinowicz, W. (1979), "Reasonable beliefs", *Theory and Decision* 10, 1979, 61-81.

Rabinowicz, W. (1985), "Intuitionistic truth", *Journal of Philosophical Logic* 14, 1985, 191-228.

Rabinowicz, W. (1995), "Global Belief Revision Based on Similarities between Worlds, in Wlodek Rabinowicz and Sven Ove Hansson (eds.), *Logic for a Change, Festschrift for Sten Lindström*, Uppsala: Uppsala Prints and Preprints in Philosophy, 80-105.

Rabinowicz, W. (1996), "Stable Revision, or Is Preservation Worth Preserving?", in André Fuhrmann and Hans Rott (eds.), *Logic, Action and Information*, Berlin: Walter de Gruyter, 101-128.

Rabinowicz, W. (1998), "Crappling with the Centipede – Defence of Backward Induction in BI-terminating Games", *Economics and Philosophy* 14, 95-126.

Rabinowicz, W. and Lindström, S. (1994), "How to Model Belief Revision", in Dag Prawitz and Dag Westerstahl, *Papers in Logic, Methodology and Philosophy of Science*, Dordrecht: Kluwer, 69-84.

Rabinowicz, W. and Segerberg, K. (1994), "Actual Truth, Possible Knowledge", *Topoi* 13, 101-115.

20
R. Ramanujam

Professor

Institute of Mathematical Sciences

Chennai, India

---

1. Why were you initially drawn to epistemic logic?

I first heard of epistemic logic in a talk by *Rohit Parikh* in 1984 ([Pa84]) and I was immediately hooked. He talked of the sum and product puzzle and my first thoughts were all about solving the puzzle rather than any logicising in relation to it. However, once I started trying to formalize the reasoning, I could see how attractive the knowledge abstraction was. Those days I was learning logic programming, so I could easily formulate the puzzle as a search problem which could be written as a Prolog program, but then all knowledge being used was hard-wired into the program. I recall a discussion with Parikh about how we should really read the knowledge assertions (by Mr Sum and Ms Product) in the puzzle as information theoretic statements rather than cognitive ones. We then started talking about what it could mean to know $f(x, y)$ without knowing x and y.

A crucial aspect of these puzzles is the use of iterative reasoning: assertions of the form A knows that B knows that something is true etc. But it was easy to see that this was also about **communication**: A knows that B does not know p, but also knows that after she utters m, B would know p, and hence informs B, and so on. I had just then started working on semantics of concurrent and communicating processes, so epistemic logic came across as being a very attractive way of discussing these things. (This is as opposed to Hoare logic and logic programming; I had studied some Hoare logics for concurrency and was exploring concurrency in logic programs then.)

2. What example(s) from your work, or work of others, illustrates the relevance of epistemic logic?

The relevance of epistemic logic should perhaps be seen from a variety of viewpoints, that of distributed computing, artificial intelligence, game theory, cryptography, ... I believe this is not merely a superficial relevance, in the sense of their dealing with knowledge and belief explicitly. The point is, I think we cannot hope to give logical / theoretical foundations to these areas without explicitly bringing in reasoning about mutual knowledge and belief. For instance, how we can codes and ciphers work without assumptions on what is common knowledge of common ignorance, and reliance on mutual knowledge between principals in the context of general ignorance ? I am not sure that work on modal epistemic logics has indeed managed to offer such foundations, but epistemic logics as such seem to me to be the best bet yet.

There are many examples that show epistemic logic at work, but it is not clear to me that they establish relevance of the logic itself, as demanded by the question. My standard would be: a result relating to a question arising from outside logic (or logical formalizations) where the answer either could not be given without using epistemic logic, or indeed if it could, the answer given using epistemic logic was unambiguously better. Such results would surely establish relevance.

I do not know if such examples can be given for cryptography, game theory, artificial intelligence etc (though I suspect they exist). In the context of distributed computing, I can single out a couple of them.

1. - [KP]: It describes a connection between knowledge as studied in mathematical economics and in distributed computing. In particular, it looks at how a sequence of communications can lead to consensus, in terms of the successive knowledge changes induced. It is very difficult (though perhaps not impossible) to see how such a result can be proved without the knowedge abstraction.
   - [DM]: It derives an optimal algorithm for consensus that tolerates crash failures, and the use of explicit reasoning about knowledge is critical. This is a dramatic example because all other algorithms take $f+1$ rounds where f is the *bound* on the number of processes that may fail during any computation, whereas the knowledge based algorithm stops as early as possible and uses

$t+1$ rounds where t is the *actual* number of processes that fail during the computation. The fact that epistemic logic led to the discovery of an optimal algorithm for a problem that was being studied by a large research community should attest to its relevance very clearly.

Once again, these are illustrative rather than being representative. Examples can (and perhaps should) be given from other domains as well.

One can also offer *mathematical* relevance: epistemic logic offers a way of looking at mathematical structures that we know well but from a new and interesting viewpoint. There are many examples of such relevance, here is a subjective selection:

1.
    - [DMP]: The paper proposes **Topologic**, a way of looking at knowledge as neighbourhoods so that we can study relationships between what we know and the effort we need to put in for knowledge. This work, and that of K. Georgatos, shows how topological structures can be derived in this fashion.
    - [KR]: The paper studies a **knowledge – time duality**: there is a *back and forth* relationship between temporal structures presented as partial orders and knowledge change presented as transition systems. In a categorical sense it sets up an adjunction; this is of relevance in the context of recent results such as [BGP].

3. What is the proper role of epistemic logic in relation to other disciplines, for instance mainstream epistemology, game theory, computer sciences or linguistics?

Ideally, I would like the role to be such as the above, in the discovery of new algorithms (in computer science), and in offering connections hard to pin down otherwise. But even without such discovery aspects, I think epistemic logic can provide a **foundational** role, helping to make precise assumptions that generally tend to be taken for granted, and in the process, re-examine these assumptions, leading to new questions and discoveries. This is especially clear in game theory and computer science where underlying models of interaction tend to be taken for granted, as also an epistemic infrastructure that is interwoven into these interaction models but made use of in reasoning abstractly about the game or process. I think raising new questions for the community can be a very valuable role for epistemic logic in these disciplines.

4. Which topics and/or contributions should have had more attention in late 20th century epistemic logic?

The single most important topic I can think of is the connection between theory of computation and epistemic logic. While epistemic logic tends to be largely descriptive, talking of A knows that p holds, questions of how A could acquire and maintain such knowledge, what resources this entails (say, in terms of memory), whether there can be a trade-off between communication and storage for knowledge (as for instance what we do these days, resorting to Google rather than store information), and so on, are important and relevant. In my opinion, the development of epistemic logic would have been much richer for attention to these questions.

A Büchi type theorem is an illustration of the kind of connection I am talking about. Büchi's theorem shows that the models of the monadic second order theory of linear orders correspond precisely to those recognizable by a finite state device. In a sense, this shows that bounded memory corresponds to definability in this logic. A similar correspondence for an epistemic logic would highlight the memory underlying epistemic reasoning in the logic. [1]

In general, developing computational foundations for epistemic theories requires more detailed answers than we have now. Rather than treating knowledge as atomic, and talking of algorithmic knowledge, explicit or implicit knowledge, distributed knowledge as types built on this notion and adding structure, it is worthwhile to consider how knowledge is structured by mechanisms that generate, store and maintain knowledge.

Let me explain this remark from one particular standpoint, that of structurally generated epistemic models: in Hintikka style models for epistemic logics, epistemic uncertainty is given by equivalence relations on worlds associated with agents. [2] However, when worlds are structured, and the model includes more details of the agents, the uncertainty relations are determined by the model structure as well. For example, consider a situation where the agents are geographically separated and communicate only by explicit exchange of messages. Then all information an agent has is, by default, local to that agent. One way of modelling this is to in-

---

[1] A class of automata is defined in [MR] and Kleene-type theorems are given, but we have a long way to go yet, as shown by the absence of corresponding algebraic theories.

[2] Often they are not equivalences but some weaker relations; the remarks here apply for them as well.

sist that the world is structured as an n-tuple $w = (w_1, w_2, \ldots, w_n)$ where n is the number of agents in the system. How communication changes informational states of agents is then given by local accessibility relations for agents. The uncertainty relation is then naturally given: $w \sim_i w'$ if $w_i = w'_i$. (There can be many other ways by which the epistemic uncertainty relations may be generated, this is merely by way of example.) Such a presentation explicates how global information is rendered in terms of local information states.

It is then a natural question to ask, given a communication model, when are standard epistemic models decomposable into local structured models ? The recent work of [BGP] is an example, but at a generic level (without communication structure). Decomposition theorems of this kind are generally hard to come by, but can add a great deal of insight to our understanding. The dependence of decomposition on the communication medium can be quite critical. (A supplementary but important question in the case of finite models is that of complexity: is there a blow-up involved in going from global to local models ?)

I also think that Parikh's work on Levels of Knowledge [Pa86], as well as Chandy and Misra's early work [CM] deserved more attention. These are paradigm exercises for what is called protocol synthesis these days. Given an initial state of knowledge of agents and a goal state, the problem is to synthesize a sequence of communications that take us from the former to the latter. These papers started a very interesting line of work in this direction, with good answers in specific models. Unfortunately, the field moved away from such analysis. But I am happy to note that the problem is receiving some attention now, and am hopeful of progress.

Another topic that has not received much attention is the connection with **program semantics**. This connection is important, if we ever hope to use knowledge abstractions directly, syntactically, inside programs and protocols. The work on knowledge based programs by Halpern, van der Meyden, Moses and Vardi (in several papers, see [Ron] for instance) offered some hope but this has not been much realised. Shoham's project on *rational programming*, using programs with utilities, was also another attempt. In fact, whether message semantics is similar to program semantics is itself an interesting question, as for instance discussed in [PR03]. But what is needed is a formal semantics of how knowledge interacts with computational elements and in my opinion, this has been lacking.

I have already mentioned cryptography. In general, the connection between security theory and epistemic logic is well known at an intuitive level, but there are surprisingly few (relatively speaking) formal results that go back and forth. The notion of *zero knowledge proofs* is a good example of this.

(I am confining myself to topics related to computer science.)

5. What are the most important open problems in epistemic logic and what are the prospects for progress?

I can think of several. Offering knowledge theoretic foundations for cryptography, games of partial information, large distributed systems (such as the Internet), can be pinned down into a set of problems. I am convinced that pursuing these would enrich these areas as well as epistemic logic.

For instance, cryptography is based on **one-way functions with trapdoor** which are easy to apply, but hard to invert in the absence of an appropriate key. Clearly, the terms *easy* and *hard* are used in an epistemic sense here. If we consider this a little further, we are led to knowledge of processes and knowledge of data, which are ontologically different from propositional knowledge of the kind typically studied in epistemic logic. One could *idealize* knowledge of data to propositional knowledge but this can lead to subtle problems, as evidenced by BAN logic ([BAN]). There have been several attempts at knowledge based modelling of security protocols such as for instance [HP], [RS05], [BRS], but these build knowledge on top of cryptographic primitives rather than providing epistemic foundations for cryptographic protocols as discussed here. Some of the challenges involved are discussed in [RS10].

Games of partial information (as also imperfect information) are typically modelled by game trees with equivalence relations on nodes associated with each player. The idea is that a player cannot distinguish between games positions that are in the same information partition. Thus they provide natural models of epistemic logics. However, such modelling can miss important foundational structure: notions of rationality underlying solution concepts are justified in epistemic terms. Stable player behaviour depends on players' extent of knowledge of the game, preferences, available strategies etc but also on players' beliefs about other players' skills, preferences and strategy selection. An atomic and unstructured notion of knowledge and agency (as modelled by the agent-indexed equivalence relation) is insufficient if our aim is to offer new epistemically based solution concepts. As we incorporate finer structure into local information of agents and into message structure, we can reason more competently about the kind of intersubjectivity in games that we discussed above. Indeed, communication structure can make the difference between undecidability and decidability of computing equilibria (as for instance in [RSim]).

Large distributed systems such as the Internet highlight another problem for epistemic logics. These are systems where agents interact with other agents knowing very little about each other, do not even know how many agents there are in the system, how others may behave in given situations. And yet, knowledge is distributed in these systems, a flexible model of interaction and communication provides for mutual introduction and information exchange within well-defined norms. What are the properties of such distributed knowledge ? How much intersubjectivity is needed for system stability ? What predictions are possible for the long range behaviour of such systems ? Agent types are important in such systems and mutual knowledge of types can be significant for their analysis (for an example involving imitation types, see [PR10]).

Then there are foundational problems relating to epistemic logic itself. For instance, the question of **agency** and **identity**: typically epistemic logics treat identities as rigid elements, like in A knows that B knows that p holds. But surely identity is in turn determined by capabilities and knowledge, and working this out is an important open problem, especially in the analysis of systems with large scale interactions (such as the Internet again, where identity is fluid and constructed).

The relation between knowledge and **reasoning** is critical as well. Epistemic logics tend to take a reasoning infrastructure for granted and build knowledge on top of it, often without changing the underlying reasoning very much. But very little reflection is needed to see that knowledge and reasoning depend on each other and transform each other, but formalizing this is quite challenging.

I have mentioned the interdependence of knowledge and **communication** several times above, and I see this as the most immediate arena in which prospect for progress can be seen. This arena intersects game theory, computer science, distributed computing, linguistics, cognitive science and information theory in its generality, and the advent of techniques from all these disciplines focussing on questions of relevance to epistemic logic offers great promise.

20.1 REFERENCES

[BRS] A. Baskar, R. Ramanujam, and S.P. Suresh, Knowledge-based modelling of voting protocols, *Proc. TARK*, 62–71, 2007.

[BGP] J. van Benthem, J. Gerbrandy, and E. Pacuit, Merging frameworks for interaction: DEL and ETL, *Proc. TARK*, 72-81, 2007.

[BAN] Michael Burrows, Martin Abadi, and Roger M. Needham, A logic of authentication. *ACM Transactions on Computer Systems*, 8(1):18–36, Feb 1990.

[CM] M. Chandy and J. Misra, How processes learn, *Proc. PODC*, 204-214, 1985.

[DMP] A. Dabrowski, L. Moss and R. Parikh, Topological reasoning and the Logic of knowledge, *Ann. Pure Appl. Logic*, vol 78, 1-3: 73-110, 1996.

[DM] C. Dwork and Y. Moses, Knowledge and common knowledge in a distributed environment: Crash failures, *Information and Computation*, vol 88, number 2, 156-186, 1990.

[HP] J. Y. Halpern and R. Pucella, Modeling adversaries in a logic for security protocol analysis, *Proc. Formal Aspects of Security*, 115-132, 2002.

[KP] P. Krasucki and R. Parikh, Communication, consensus and knowledge, *Journal of Economic Theory*, vol 52, 178-189, 1990.

[KR] P. Krasucki and R. Ramanujam, Knowledge and the ordering of events in distributed systems, *Proc. TARK*, 267-283, 1994.

[Ron] Ron van der Meyden, Knowledge based programs: on the complexity of perfect recall in finite environments, *Proc. TARK*, 31-49, 1996.

[MR] S. Mohalik and R. Ramanujam, Automata for epistemic temporal logic with synchronous communication, to appear in *JOLLI*, 2010.

[Pa84] R. Parikh, Logics of Knowledge, Games and Dynamic Logic, *Proc. FST&TCS*, Springer LNCS 181, 202-222, 1984.

[Pa86] R. Parikh, Levels of Knowledge in distributed computing, *Proc. LICS*, 314-321, 1986. (An extended version, jointly with P. Krasucki, appeared in *Sadhana*, 1989.)

[PR03] R. Parikh and R. Ramanujam, A Knowledge Based Semantics of Messages, *JOLLI*, 12(4), 453-467, 2003.

[PR10] S. Paul and R. Ramanujam, Imitation types in large games, *Proc. GANDALF*, 2010.

[RSim] R. Ramanujam and Sunil Simon, A communication based model for games of imperfect information, *Proc. CONCUR*, 2010.

[RS05] R. Ramanujam and S.P. Suresh, Deciding knowledge properties of security protocols, *Proc. TARK*, 219-235, 2005.

[RS10] R. Ramanujam and S.P. Suresh, Challenges for epistemic theory from security protocols, to appear in *Logic at the Crossroads*, (Ed. A. Gupta and J. van Benthem), Synthese Library Series, 2010.

21

Krister Segerberg

Professor Emeritus

Uppsala University, Sweden

A Conversation about Epistemic Logic[1]

Dramatis personæ: SANDERSON *(the Professor's teaching assistant);* JOHN *and* PHOEBE *(philosophy students);* SEBASTIAN *and* CHARLOTTA LINNÉA *(the Professor's grand-children aged seven and four, respectively). The scene is the Professor's office, moderately untidy. On one wall a small blackboard. Three people in the room: Sanderson, Sebastian and Charlotta Linnéa. They have been waiting for John and Phoebe, who suddenly enter the room.*

JOHN. Hi, we have come from the student newspaper to interview the Professor. I am John the journalist.

PHOEBE. And I am Phoebe the photographer.

SANDERSON. Hi, and welcome. I am Thomas Anderson but my friends call me Sanderson. I am the Professor's T.A. and B.S. B.S. meaning baby sitter. Meet Sebastian and Charlotta Linnéa, the Professor's grand-children.

PHOEBE. *(Smiling at the children.)* Hi!

CHARLOTTA LINNÉA. I am not a baby!

SANDERSON. Of course not. I am sorry the Professor is late, but we are here to answer your questions.

JOHN. And the children?

---

[1] Reprinted from *Epistemology: 5 Questions*, edited by Vincent F. Hendricks and Duncan Pritchard. New York, London: Automatic Press / VIP, 2008: 283–304.

SANDERSON. Actually, the Professor suggested to begin with them.

JOHN. I know the Professor is eccentric, but this is ridiculous. What do children know?

CHARLOTTA LINNÉA. (*Proudly, not waiting for an invitation to speak.*) I know how old I am.

PHOEBE. (*Friendly.*) And how old is that?

> *Charlotta Linnéa silently displays one hand: fingers spread out, thumb hidden.*

JOHN. (*Not so friendly.*) And just how much is that?

CHARLOTTA LINNÉA. (*Again displays her hand.*)

JOHN. So you don't know?

CHARLOTTA LINNÉA. Four, silly!

JOHN. So you *know* you are four, do you?

CHARLOTTA LINNÉA. (*Nods importantly.*)

JOHN. And how do you know?

CHARLOTTA LINNÉA. (*Counts on her fingers.*) One, two, three, four.

PHOEBE. (*Supportive.*) There you are! She really knows that she is four. Very clever!

JOHN. No, I don't think so. She has been told she is four and just repeats that. Counting to four does not mean you know what four is. I am not even sure that what she did was counting—a parrot could be taught to do what she did.

CHARLOTTA LINNÉA. (*Indignant.*) I am not a parrot!

JOHN. So how much is five plus seven?

PHOEBE. (*Trying to be helpful.*) That is too difficult for a little girl. Let's ask Sebastian instead. How old are you, Sebastian?

SEBASTIAN. Seven. And I can count to one thousand.

PHOEBE. You are older than your sister.

SEBASTIAN. Three years older.

JOHN. How do you know?

SEBASTIAN. Seven minus four is three.

JOHN. You *know* that?

SEBASTIAN. I can show you. But I need some coins.

> *At the request of Phoebe, John produces seven coins—as it happens, four nickels and three quarters. Sebastian begins by counting them.*

SEBASTIAN. One—two—three—four—five—six—seven.

> *Sebastian then counts the nickels, handing each back to John as he is counting.*

SEBASTIAN. One—two—three—four.

> *Sebastian finally counts the remaining coins—the quarters.*

SEBASTIAN. One—two—three. So the answer is three.

> *Sebastian pockets the three coins.*

JOHN. Hey, that is my money!

SEBASTIAN. But it is just three quarters!

CHARLOTTA LINNÉA. He is mean!

JOHN. This is getting tiresome. Give me back my money!

PHOEBE. (*Gently, as if to herself.*) "May I *please* have my money back!"

> *Sebastian returns the money.*

SEBASTIAN. I'd settle for one quarter.

SANDERSON. Children: Why don't you go and do your home-work in the library. After that we will all go to Starbucks.

SEBASTIAN. But weren't we supposed to tell them about knowledge? That is what Grandpa said.

SANDERSON. But you have! Thanks, and see you later!

SEBASTIAN. I'd rather go to McDonald's.

CHARLOTTA LINNÉA. He's mean! I am not a parrot.

The children depart.

*

SANDERSON. So where were we? You wanted to know about epistemic logic?

JOHN. Yes, we are running a series in the campus newspaper about what's hot in various disciplines.

SANDERSON. And you were told that epistemic logic is "hot"?

JOHN. Right. So now we'd like to know, what has knowledge got to do with epistemic logic?

PHOEBE. Or, rather, what has epistemic logic got to do with knowledge?

SANDERSON. O.K., so let us start by saying that there are lots of important concepts involved here: not only knowledge but also belief, information, values, and action, to mention some. They are all intertwined.

PHOEBE. Aren't "belief" and "information" two words for the same thing?

SANDERSON. No, information is something you get and which you have to evaluate: whether to accept it or not. It may lead to belief, even knowledge. Or it may not. Now, what the Old Fog always stresses is that there are several dimensions to modelling knowledge: presentation, representation and commitment.

JOHN. What on earth is that supposed to mean?

PHOEBE. Who is the Old Fog?

SANDERSON. The Old Fog is the Professor. When we really want an answer, he always says that he hasn't got the foggiest idea. He is a dear, Old Foggie, but not always easy to understand. Please don't tell the children that we call him that! He is a logician, you know. Or a kind of logician. Model theorist. For him, models is everything. They represent something. They pack information. They help you see things. But just having a model is not enough, you must also know what to do with it. To be useful, models must be applied. Like maps.

JOHN. Maps?

SANDERSON. Maps. Remember what Ramsey is supposed to have said ...

PHOEBE. (*Enthusiastically.*) Oh, I know that!

JOHN. Ramsey?

PHOEBE. "Beliefs are maps we steer by." It's beautiful!

SANDERSON. And true in a very deep sense. You want to go from point A to point B. You get a map, or you get a description of how to get from A to B. Instructions or perhaps a set of rules. Whatever the form of the information, you hope that it correctly represents the features of the world that are of interest to you. In this case, to get from A to B.

PHOEBE. So maps are examples of models?

SANDERSON. Yes. I only wish there were a better term. "Model" is already used in so many contexts. "Map" would be misleading—too special. To get something neutral, how about "m.a.p."?

JOHN. I hate acronyms.

PHOEBE. What would "m.a.p." stand for?

SANDERSON. Well, for m read "multiple-use" or "multi-purpose" or something like that. For a, try "auxiliary" or "action-guiding". And for p, read "presentation". M.a.p. One m.a.p., many m.a.p.s. Not pretty, but at least a neutral term.

JOHN. For what?

SANDERSON. To cover all the concepts mentioned above: model, map, algorithm, set of rules, set of instructions, what not. I want to leave it vague.

JOHN. In other words, "m.a.p." is an ugly term for we don't know what.

SANDERSON. Sure, if you want to put it that way.

PHOEBE. May we have some examples?

JOHN. No, wait, I cannot stand "m.a.p.". If you don't like "model", how about "muddle"?

PHOEBE. Look, this is silly. Let us just stick with "model" and agree to remember that the term can cover lots of different things. The term would be clarified if we could have some examples?

SANDERSON. I am willing to compromise: *model, O.K.? You can pronounce that "star model", if you want. But perhaps best to let the star be silent. Like a nonstandard model. It is actually a nice term. Thanks for your input!

JOHN. (*Without enthusiasm.*) You are welcome.

PHOEBE. I take it that a model in the ordinary sense would also – or could also – be a *model. But what other examples are there?

SANDERSON. All right. Here is a map of New Zealand. A lot of symbols and a number of names. That is what I called a presentation: just an abstract structure, really. But there is a way to read it. For example, this dot represents Auckland, this dot Wellington, this line is the main road between them. That would be the representation.

JOHN. The re-presentation, as Heidegger would say.

SANDERSON. Let us keep Heidegger out of this, shall we?

PHOEBE. And commitment?

SANDERSON. You would not want to use the map if you didn't think it was true to the facts. Say you have several maps to choose between and that they differ in important ways. Then you have to make a choice. In my jargon: to make a commitment.

PHOEBE. Which then can be the basis for further action. Like motoring from Auckland to Wellington.

JOHN. (*Scornfully.*) "Motoring"!

PHOEBE. Isn't that what they say in New Zealand?

JOHN. I'm not sure they ever did. Today they certainly don't.

SANDERSON. What the children did can be seen in the same light. Lotta-Linnéa has a procedure of ticking off her fingers; at the same time she associates them with numbers; and then she assures you that this is the way to get an answer to your question. Similarly, Seb arranges his coins and performs an operation on them. Then he places an interpretation on what he has done. And finally, he gives you his answer. So in both cases we can use *models to represent what happened.

PHOEBE. Let us see if I got this. There are several parts to your modelling. One is an abstract structure of one sort or another that you call the *presentation*; one is an interpretation of the presentation which you call the *representation*. Then there is an *agent* and a relationship between agent and presentation-cum-representation you called the *commitment*. A bit complicated, but with a little good will I think one can accept that.

JOHN. It certainly requires more good will than I have.

SANDERSON. Let me add two things. One is that the representation really contains two elements: a *language* appropriate for the presentation, and an *interpretation* of that language over the presentation.

PHOEBE. Which means that if we want to think in terms of the usual jargon we can think of a *model as a triple $\langle \mathcal{P}, \mathcal{L}, \mathcal{I} \rangle$, where $\mathcal{P}, \mathcal{L}$ and $\mathcal{I}$ are, respectively, the presentation, the language and the interpretation. But that still leaves out the commitment part.

JOHN. I should have warned you that Phoebe is a philosophy major. She is good at talking like that. Me you can disregard, I'm majoring in English.

SANDERSON. I see. Well, it is all right if you want to analyse the *model as you do. The way you put it, we are reminded of the models that logicians use. For some purposes that might be too demanding. Too exact. Up to a point it is fine. But there may be differences. For example, the presentation $\mathcal{P}$ may be vague, and the language $\mathcal{L}$ may be very restricted, perhaps just a fragment of a language.

But even more important – this is the second thing I wanted to add – it must be clear that the commitment part is a totally different kind of thing. It belongs to pragmatics, if you know what I mean. The commitment, like the agent, is outside the *model. Let me draw you a picture.

> *Goes to the blackboard. First draws a circle with the label $\mathfrak{M}$ attached to it. Then draws a little man outside the circle and a dotted line connecting him and the circle.*

JOHN. Great piece of art. What is it supposed to mean?

SANDERSON. The little man is the agent. It could be you, if you wish. (*Writes "John" under the little man.*) The dotted line represents your commitment to $\mathfrak{M}$. And $\mathfrak{M}$ is simply the *-model that represents the content to which you are committing yourself. The important thing is that you are literally outside the *model. The commitment is outside $\mathfrak{M}$—it belongs to pragmatics.

PHOEBE. So this figure illustrates "John knows that p", say. But suppose that I know that John knows p and I want to illustrate that. Then I need another circle?

SANDERSON. Yes.

> *Draws a big circle around all of the old figure on the blackboard and places a little woman under it with the name "Phoebe" attached.*

PHOEBE. And then I should think of a new, more inclusive language appropriate for the more inclusive presentation. That is, I really need a bigger *model, $\mathfrak{M}'$ say.

SANDERSON. Correct. (*Attaches the label "$\mathfrak{M}'$" to the second circle.*) And notice that you are committing yourself to $\mathfrak{M}'$ but not to $\mathfrak{M}$.

PHOEBE. Although I could commit myself to $\mathfrak{M}$ as well, if I wanted to?

SANDERSON. Of course, but it would be an extra. And if I go on to say that Phoebe knows that John knows that p, then in effect I am adding yet another circle.

> *Draws a third circle around everything, attaching the label $\mathfrak{M}''$ to it, and under it draws a little man with the label "Sanderson" attached to it.*

PHOEBE. I understand. But then it occurs to me the relationship between agent and *model can be of different kinds. This is something I had wanted to ask you about anyway. Surely there can be different kinds of relationships between agent and *model? Endorsement? Leaning towards? Liking? In short, having some kind of attitude?

JOHN. Like a propositional attitude? Isn't that a term you philosophers use?

SANDERSON. Actually I have thought of the relationship as the result of some action. Commitment and endorsements are actions, you know. But perhaps it is enough to have an attitude. I shall have to think about it. But thanks for the suggestion!

JOHN. Now what has all this got to do with knowledge?

SANDERSON. O.K., what does it mean to say that you know something? According to the idea that I have in mind it means that if you say something is true, then it need not be true but it has to be true in the *model that is implicit—the one to which you are committed.

JOHN. On your map the cities are red, the country is green, and the roads are black. So that is what you are committed to? New Zealand is green?

PHOEBE. That's silly, John!

SANDERSON. Well, you know, New Zealand *is* green! Seriously, there is an important point here: we are only allowed to view the presentation through the lens of the representation. That is why the language of the *model is important.

JOHN. The *language, perhaps. You see, I am with you!

SANDERSON. And that language – the *language, if you wish – is usually quite restricted. In the case of maps, if colours are not just merely decorative, one needs a code to interpret them.

PHOEBE. O.K., let me try to express this more formally, Would it be right to put this idea in the following way: a knows that p if and only if there is a *model $\mathfrak{M}$ such that

- a is committed to $\mathfrak{M}$,

    there is a formula ϕ of the language of $\mathfrak{M}$—the *language, to use John's term—such that $p = [[\phi]]_{\mathfrak{M}}$,

- $\mathfrak{M} \vDash \phi$.

SANDERSON. Here I suppose that $p = [[\phi]]_{\mathfrak{M}}$ means that what p means is expressed in $\mathfrak{M}$ by ϕ? And that $\mathfrak{M} \vDash \phi$ means that ϕ is true in $\mathfrak{M}$?

PHOEBE. Right. And if you replace "a is committed to $\mathfrak{M}$" by "a endorses $\mathfrak{M}$" or "a supports $\mathfrak{M}$" you get belief rather than knowledge?

SANDERSON. If we suppose that commitment is stronger than endorsement and support. Commitment should be maximal in some sense.

JOHN. Like you would be willing to bet your life on it?

SANDERSON. No, I don't want to get into betting ratios and all that. In fact, I am not sure knowledge should be seen as the limiting case of belief. Beliefs are easily formed. For the most part they are revisable. Some have hardened into knowledge. That is how I see knowledge: belief that is no longer revisable. Or, rather, belief that a rational agent would not think of revising.

JOHN. (*Ironically.*) Unless he really, really had to.

PHOEBE. Are you telling us that there is no real knowledge?

SANDERSON. What do you mean, "real knowledge"?

PHOEBE. Well, you know, real knowledge. Knowledge that must be right. Knowledge that will never need to be revised. Knowledge that cannot be revised. At least not by a reasonable person.

SANDERSON. So what is a reasonable person?

PHOEBE. Well, a rational person.

JOHN. So what about mathematical knowledge?

SANDERSON. (*On the defensive.*) Mathematical knowledge?

*

JOHN. Yes, mathematical knowledge. Most of us think that there is something special about mathematical knowledge: it is true—necessarily true. It is certain—absolutely certain. Is that just prejudice?

SANDERSON. Mathematical knowledge is a difficult one. But let us begin with prejudice, which is an interesting concept in its own right. Interesting also to epistemologists. Notice how you say "just prejudice"? The word "prejudice" has a bad ring to it. Many are prejudiced towards prejudice. But actually I think we need prejudice. In many situations there is not time to check everything, so you fall back on stock beliefs that you hold. And most of the time they serve you well. The fact that prejudice sometimes leads you astray is true, but it is a price you may have to pay.

PHOEBE. Non-monotonic logic? Tweety is a bird, and all that?

SANDERSON. Right. Prejudice is often like a set of default rules that you have. Jumping to a conclusion saves time. Even if you sometimes jump to the wrong conclusion, in general that is a price worth paying.

JOHN. You already said that.

SANDERSON. Forgive me, but you are here to interview me about what epistemic logic has done for knowledge, and I think of non-monotonic logic – some kinds, at least – as kinds of epistemic logic.

And Reiter's default logic is one of the most important. It does tell you something about knowledge. Or at least rational belief.

JOHN. Fine. Now tell me about mathematical knowledge.

SANDERSON. I agree that mathematical knowledge is a special case. Can we come back to that?

*

JOHN. O.K., let us go back to your theory of muddles. It smacks of relativism. Are you a relativist?

SANDERSON. In one sense, yes. Knowledge claims, like existence claims, are relative to a *model.

JOHN. I hate relativism. Does Santa Claus exist?

SANDERSON. No.

JOHN. But aren't there muddles in which he exists?

SANDERSON. Yes. But he does not exist in any *model that I am willing to endorse.

JOHN. Wouldn't your theory have made it difficult for Descartes to get started? "I think in a star muddle to which I am committed, therefore I am." It does not roll well off the tongue.

PHOEBE. Perhaps it sounds better in Latin.

JOHN. But what if you commit yourself to one muddle in which it is true that p and also commit yourself to another muddle in which it is false that p?

SANDERSON. Well, then I am contradicting myself.

JOHN. Worse than that: according to your own theory you know that p and also you know that not-p.

PHOEBE. I was going to ask about that. It is not a problem for your theory if it allows an agent to believe both p and not-p (although it may be a problem for the agent!). But to know both p and not-p does not sound right.

JOHN. Provided you agree that knowledge implies truth. Which I thought every epistemic logician would do, however crazy.

PHOEBE. In other words, do you really deny the validity of the T-schema?

JOHN. Excuse me, but what it the T-schema?

PHOEBE. The T-schema is the corner stone of epistemic logic, in symbols $\mathbf{K}\phi \to \phi$. In English: if you know that ϕ, then it is true that ϕ.

JOHN. You call that English?

SANDERSON. Let us see how I can explain this. In order to discuss formal logic, you need *models of a certain kind. In particular, the chosen *language must contain the operators you wish to study. You can certainly find *models that validate the T-schema, just as you can find *models that don't. In either case you may commit to the *model in question. But no-one, not even you, can think about this without going to a bigger *model that sort of engulfs the original *model. Moreover, depending on your conception of knowledge, you commit to the bigger *model only if the agent's attitude in the smaller *model is the right one. So to speak.

PHOEBE. Could you go over that again?

JOHN. Yeah, I couldn't follow him either.

SANDERSON. First we have a *model, $\mathfrak{M}$ say, with respect to which some statement p is true (meaning that $\mathfrak{M} \vDash \phi$, for some formula ϕ in the language of $\mathfrak{M}$ corresponding to the informal p). Then we want to bring knowledge into the picture. This means introducing an agent, a say, who has a certain attitude towards $\mathfrak{M}$. Now that we are considering a and his relšation to $\mathfrak{M}$ we have in fact introduced yet another *model, containing (representations of) a and $\mathfrak{M}$. Call this new *model $\mathfrak{M}'$. We will obviously be making a number of assumptions. For example, in a simple case we will assume that the language of $\mathfrak{M}$ is included in the language of $\mathfrak{M}'$, and we will take it that $\mathfrak{M}$ and $\mathfrak{M}'$ agree on what is common to the two languages. Most important in this context, if you are taking a Cartesian view of knowledge ...

JOHN. (*Interrupting.*) What is Cartesian knowledge?

SANDERSON. Knowledge that is indubitable and absolutely certain. As I was saying, if you are taking a Cartesian view of knowledge, then you must assume that the agent in $\mathfrak{M}'$ will not accept anything that can be said in the common language that is not true in $\mathfrak{M}$.

JOHN. What if $\mathfrak{M}$ got it wrong, then we would still say the agent knew everything in $\mathfrak{M}$? Including the falsehoods?

SANDERSON. We would say that he "knew" them. That he thought he knew them.

PHOEBE. So if I understand you, the validity of the T-schema is built into your modelling. That is, when it is valid in a *model it is because you have built the model in such a way as to validate it.

JOHN. I may not be as smart as Phoebe, but is seems to me that instead of talking about the commitment of the agent you should be talking about, say, the *relation between agent and muddle. You decide what the relation is and then you, the analyst, evaluate it. You decide whether the agent knows something or not.

SANDERSON. You, the analyst, draw the picture. Every time it is the little man outside all the circles that draws the entire diagram.

PHOEBE. Or woman.

SANDERSON. The important thing is that the analyst is always on the outside.

JOHN. The pot of gold at the end of the rainbow!

SANDERSON. Let me just say, before moving on to networks of *models, that there are many topics to discuss in connexion with even a single *model. As far as I am concerned, the most difficult of them is getting knowing-how into the picture.

JOHN. To get know-how into the picture?

SANDERSON. Not know-how: knowing-how. There is a difference between knowing-that and knowing-how: knowing that something is the case, and knowing how to do something. The Germans have different words for this: the *wissen-können* stuff, you know.

PHOEBE. Knowing-that and knowing-how? That's interesting. How about knowing-who, knowing-when, knowing-why? Perhaps there are other concepts like that?

SANDERSON. I think the idea is that if you can handle knowing-that and knowing-how, you should be able to take care of the others as well. Anyway, as I have said before, knowledge and belief are very closely connected with action. So here we are leaving out a big chunk.

JOHN. Good, we have not got all day. Was there anything else?

*

SANDERSON. If we are interested in epistemic agents, if I may call them that, then the concept of a *model is only a beginning. Here is how I see our agenda:

§1 Defining '*model'.

§2 Defining 'network of *models'.

§3 Studying how networks of *models change over time.

The first point, §1, we have already dealt with. Of course there are numerous other things to discuss but, as John points out, time is limited. Let us move on to §2: networks.

An agent has a lot of representations of those aspects of the world that are of interest to him. You may think of this as a craftsman having lots of tools. The *models we have talked may be interrelated. I can think of three basic relationships, although there may be others: subsumption, refinement, and allotropy. Again, maps provide an excellent example. If you cut out a piece of a bigger map, you may be said to have a *submap* of the larger map.

PHOEBE. So it makes sense to talk about "sub-*models", even though the term is awkward. Perhaps "*submodels" is better. The *model $\mathfrak{M}$ we just talked about is a *submodel of the bigger *model $\mathfrak{M}'$.

SANDERSON. Sometimes a map of a country will contain maps of the bigger cities of that country which appear like filled cirles or something on the main map; those smaller maps may be said to be *(partial) refinements* of the big map. Finally, you may map out different aspects of one and the same object. For example, you may have lots of different maps of the same city: a subway map, a street map, a geological map, and so on. In this case we use the term *allotropy*: those maps are all allotropes.

JOHN. "Allotropy" is a term chemists use for the property of a substance to exist in more than one form.

SANDERSON. That is right. Anyway, to make a long story short, in order to represent an agent's knowledge along these lines, we should think of a big network of *models interrelated in the ways mentioned. And perhaps others!

PHOEBE. Perhaps we should also be thinking of what the agent can do with these *models. Perhaps this is another place where the distinction between *wissen* and *können* comes in. At a higher level this time.

*

JOHN. Speaking of time, I'm not sure I can stay much longer.

PHOEBE. Speaking of time, shouldn't we talk also about change? Wasn't that §3 on your agenda?

SANDERSON. Of course, change is important. If we want to model the knowledge of an agent, we must consider change. A network of *models can only give the epistemic state of an agent at a particular time. But the epistemic state may change at any time. There are changes of at least two kinds: change due to new information and change due to introspection.

JOHN. It is obvious that we process new information almost all the time. But introspection?

SANDERSON. Introspection also goes on almost all the time. We reflect about what we know or believe. We see new connexions. We discover contradictions. We continue to work on *models we have already accepted.

PHOEBE. Trying to clean up, as it were.

SANDERSON. Suppose you review your current network of *models. Suddenly you realize that there is an implicit contradiction in the network. Obviously you try to revise it to get rid of the inconsistency. Or suddenly you see a connexion between two or more *models that sort of fuse into one. Or you add a new *model that is somehow inspired by the old one.

PHOEBE. Like set theoretical union?

SANDERSON. Union is one example, but union is more than union, if you know what I mean.

JOHN. Union is more than union?

SANDERSON. Yes: there may be something creative about the new *model that goes beyond the old ones. The new *model may be greater than the sum of the old ones. Call it "fusion". It has to do with getting an insight! Realizing something you knew without knowing you knew!

JOHN. You know, I sort of like this. I know without knowing, and the fusion is greater than the fuses. Confusion is greater than Confus-ius. It has a nice rhythm to it.

SANDERSON. Confus-ius? Do you mean Confucius? I don't understand.

PHOEBE. Don't pay any attention to John, Mr Anderson, he is just being silly. What I would like to know is this: Isn't change due to new information what is studied in the theory of belief revision?

SANDERSON. Yes, that has mushroomed into a big industry. But I think there is still room for more work. Personally, I am particularly interested in the concept of metaphor. Did you ever notice that new information can come in the form of a metaphor?

PHOEBE. I am sure metaphor has other uses than conveying information.

SANDERSON. No doubt, but conveying information is one of them. And when successful, it can be a very economic way of communication. Metaphor is of course intimately connected with change. Here knowledge change and language change go hand in hand.

PHOEBE. Could you be more precise?

SANDERSON. Successful metaphors function as a kind of inference rules. Suppose a poet says, "Oh, the pain when bud bursts into flower!".

JOHN. That is not even a sentence.

SANDERSON. That's right. Not a declarative one, at any rate.

JOHN. And it does not convey new knowledge. In fact, it is false: flowers don't feel anything.

SANDERSON. I suppose not. But in certain situations or contexts even a metaphor like that one may give you new knowledge—or perhaps we should call it insight or information. I can suggest a number of possible contexts in which the poet's outburst can make you draw an inference.

PHOEBE. Non-monotonic inference?

SANDERSON. Of course. An inference rule that you apply only if you want to. At your own risk and only if you want to. Take the phrase I quoted about the pain when buds burst into flowers. Here are some very simple examples.

First context: we are discussing teenagers and their problems. One possible inference is that it is painful to be a teenager. Which may be news to some. Another possible inference is that it is painful to have to deal with teenagers. Which also may be news to some.

Second context: we are discussing the economic situation today and the phenomenon of merger—smaller companies merging with or being swallowed up by bigger ones. One possible inference: that for some, economic growth can be painful.

Third context: ...

PHOEBE. (*Interrupting.*) But that is not generating new knowledge, it is rather focussing on features we already know. It makes knowledge we already have more vivid. It makes you see things, if you know what I mean.

SANDERSON. Metaphor certainly has that function as well. But I insist that metaphor can be a quick and effective method of transmitting information.

PHOEBE. But not reliable. A metaphor provides a form, you fill it with meaning.

JOHN. Now, there's a metaphor! I wonder what it means?

SANDERSON. But isn't it the same with inference rules?

PHOEBE. This is too vague for me. Shouldn't you work out a theory to back you up?

SANDERSON. I entirely agree. It would be nice if I could. Only I don't yet know how to do it. We need a topology for the space of (some class of) *models. Each metaphor inducing its own topology, you know. A metaphorical statement is usually either nonsensical or blatantly false in the *model with respect to which you are trying to evaluate it. So you need another *model to make sense. And that *model must be relevant to your concerns—as close to your current *model as possible.

PHOEBE. That is why you want topology, to express nearness?

SANDERSON. Right. But there are all sorts of problems. For example, metaphors are often incomplete in the sense that they can be cashed out in different ways. And there is in the class of metaphors a relation of subdivision or refinement that makes communication stimulating but also treacherous: the sender may have one subdivision in mind, while the receiver may seize upon another.

PHOEBE. But people have written tons about this. Do we really need epistemic logic here?

SANDERSON. Well, the theory of metaphor may not need epistemic logic, but for me as an epistemic logician it would be interesting to see whether some of the most important features of metaphor could not be analysed within the theory of belief revision. I think it would be possible.

JOHN. Why am I suddenly reminded of Dr Johnson's dancing dogs?

PHOEBE. John, that's not nice!

JOHN. (*Unperturbed.*) So what about mathematical knowledge?

SANDERSON. I wish the Professor were here to answer that one ...

JOHN. The Professor! The Old Fog! This is like waiting for Godot. I am leaving.

*

PHOEBE. (*Politely.*) I am afraid we have already stayed too long.

JOHN. We certainly have! At least I have. I thought we would be learning something about epistemic logic. But your muddles don't seem to have anything to do with anything.

PHOEBE. Don't think I haven't enjoyed our conversation. But at the same time I have to agree with John that I had expected something different. I understand that there is a lot of really important work going on in contemporary epistemic logic. The TARK conferences! The Amsterdam school! You might have told us about that.

SANDERSON. But you are philosophers! Much of what is going on in epistemic logic today might is really epistemic engineering.

JOHN AND PHOEBE. Epistemic engineering!

PHOEBE. That sounds patronizing.

JOHN. Disrespectful, I would say.

SANDERSON. No, no, I don't want to be appear either patronizing or lacking in respect. Engineering is very important. For example, it was engineering that put men on the moon, one of the most astounding feats of our time. Without the engineers we would never have got there. But just as we need philosophy of science to understand what scientists do, so we need philosophy of epistemic logic to understand what those clever people in epistemic logic do. What what they do has to do with knowledge. I mean, philosophy of soccer is not the same as soccer.

JOHN. Great! We sure need philosophy of soccer. Like we need philosophy of muddles.

PHOEBE. It is actually an interesting idea: in order to understand soccer it is not enough to play it. Not necessary, of course, but perhaps not sufficient either. Perhaps Mr Anderson is right.

JOHN. Understand soccer!

SANDERSON. I think there must something more to the philosophy of epistemic logic that just building those very technical models. That is all I am trying to say, really. (*With some emotion.*) That is what my muddled and ridiculed efforts were aimed at.

JOHN. (*Unperturbed.*) You mean "*modelled" efforts, don't you. The star silent, of course.

SANDERSON. No, I really mean "muddled". I am trying to work all this out, but as you can tell I find it difficult.

PHOEBE. I think I know what you mean. Game theory, for example. It has an important rôle in the analysis of rationality.

SANDERSON. Essential.

PHOEBE. O.K., game theory has an essential rôle in the analysis of rationality. It helps us define a kind of rationality. Or one kind of rationality. But mathematical analysis of abstract games is not the same as philosophical understanding of rationality.

SANDERSON. The theory of games is a beautiful creation. And of course I agree that game theory contributes to our understanding of rationality. In the same way I would hold that epistemic logic contributes to our understanding of knowledge and belief. I only wish I could understand more exactly how!

JOHN. I thought that was what you were supposed to tell us. That's why we came!

SANDERSON. Well, believe me or not, I have done my best. By the way, in epistemic logic I include doxastic logic.

PHOEBE. You might as well regard epistemic logic as a special case of doxastic logic.

SANDERSON. That is actually what I do, but I have common linguistic usage against me.

*

JOHN. So let's see what we have. (*Gets out a notebook and starts writing.*) You think there is a need for epistemic logic in order to analyse knowledge,

SANDERSON. Yes, I think epistemic logic has a rôle to play.

JOHN. Furthermore, you think that current epistemic logic must widen its purview, notably by paying more attention to the agent.

SANDERSON. And why knowledge is important to him.

PHOEBE. Or her.

SANDERSON. Something like that. The pragmatic aspect.

JOHN. You think that what is going on today is just epistemic engineering.

SANDERSON. That is not what I said. Epistemic engineering, as I called it, is very important. And I will tell you this: if my programme—vague and feckless as it may be ...

JOHN. What does "feckless" mean?

SANDERSON. In this context, how about "futile". If my programme were to be successful, then it, too, might develop into epistemic engineering. If it were useful, that is. If not useful, it would simply develop into another branch of mathematics.

JOHN. Not so fast, I am trying to take notes. "If not useful, it would become just mathematics."

SANDERSON. Not "just mathematics": just "mathematics". You see, there is a general pattern in abstract thinking. Ideas worth anything, if they have a minimum of structure, eventually they will be formalized. Formalisms, once born, live lives of their own. Mature formalizations grow into new disciplines. Eventually they may ossify—that happens. In rare cases they may even, in a sense, die.

JOHN. Wow! Quite a speech. So all formal philosophy ends in ossification and death?

SANDERSON. Again, that is not what I said. At least it is not what I meant to say. Perhaps all I want to say is, if you push formal philosophy too far, it ceases to be philosophy. But that isn't necessarily a bad thing. Think of the ugly duckling!

JOHN. Philosophy as the ugly duckling. Brilliant!

SANDERSON. I have a feeling you aren't really trying to understand what I am trying to say. But then I realize I am not very good at expressing what I think. I am as bad as the Professor.

JOHN. Not to worry. I am beginning to think I might be able to use some of this, after all.

*

Enter the children, Charlotta Linnéa carrying a drawing.

PHOEBE. Oh, a drawing. How nice!

CHARLOTTA LINNÉA. (*Walks up to John.*) For you!

PHOEBE. What is it? May I see? What a beautiful bird! Oh, I get it! (*Laughs.*) It is a parrot! And it has a little beard, just like John!

JOHN. (*Not amused.*) I'm out of here. Good-bye!

SANDERSON. Wait a second, what about mathematical knowledge? If you wish I could try to say something.

JOHN. Some other time! (*Prepares to leave.*)

CHARLOTTA LINNÉA. (*Importantly.*) I have a question. For you!

She points at John, clearly enjoying being the centre of attention.

CHARLOTTA LINNÉA. Why is water wet?

JOHN. I haven't got the foggiest idea.

SEBASTIAN. (*Surprised.*) That's what Grandpa always says!

John leaves.

CHARLOTTA LINNÉA. (*With evident satisfaction.*) *I* knew the answer to *my* question.

PHOEBE. (*To Charlotta Linnéa.*) I know you did! (*To Sanderson.*) I apologize for John, Mr Anderson. He is really much nicer than you might think. But Philosophy isn't his thing, you know.

SANDERSON. So I gathered. I'm not sure I want to read his article.

PHOEBE. It's clear John did not understand you. And perhaps didn't want to. But I promise to keep an eye on the article before it is published. Now, I wonder if what you are trying to say is something like this.

Knowledge and belief are philosophically central concepts. Epistemic logic has an important rôle to play in coming to a philosophical understanding of them.

SANDERSON. Indispensable.

PHOEBE. Indispensable. Nevertheless, being able to process claims to knowledge or belief isn't all there is to the philosophy of knowledge and belief. (*Pause.*) But that is obvious, isn't it?

SANDERSON. (*With some bitterness.*) That's the problem with philosophy (my problem, at least): what I don't understand is beyond my comprehension, what I understand is trivial.

PHOEBE. (*Laughing.*) You are too young to be a cynic, Mr Anderson!

SEBASTIAN. I am hungry. Let us go to McDonald's.

SANDERSON. O.K., I can take you to Starbucks.

CHARLOTTA LINNÉA. (*Points at Phoebe.*) She can come, too. She is nice!

PHOEBE. Thank you, I'd love to come. Then perhaps I can get a picture of you and your brother?

CHARLOTTA LINNÉA. And Sanderson.

PHOEBE. And Sanderson!

THE END

22

Yoav Shoham

Professor of Computer Science
Stanford University, USA

---

1. Why were you initially drawn to epistemic logic?

"Why" is hard but "when" is easy. It was in 1986, close to the end of my graduate studies in computer science at Yale. I was invited to the first TARK conference (it then stood for Theoretical Aspects of Reasoning about Knowledge; we later changed it to Theoretical Aspects of Rationality and Knowledge, reflecting the broader scope that emerged but retaining the TARK brand...). In time I became quite involved with TARK, including serving as program chair in 1994, an invited speaker in 2009, and member its board of directors for much of its existence. But like a first love, that first conference was a formative experience for me. It was wonderful to see philosophers, computer scientists, game theorists, psychologists and other sorts convene around a shared topic and a nascent theory. The beauty of the theory left a deep impression on me, as did some disturbing questions it raised. Both the beauty and the puzzles kept me explicitly actively for about a decade, during which time I published papers on knowledge, belief, and nonmonotonic belief revision. Apparently the issues continued to have a grip on me later as well, since after a hiatus of a decade, during which I devoted much of my time to non-epistemic issues at the interface of computer science and game theory, I now find myself drawn back to that material.

If I were to take a stab at the "why" question, I would mention two factors. The first is the magic of having three quite disparate fields – philosophy, game theory, and computer science – converge on the same formal notion of knowledge. Of course when you look closely at magic you see the sleight of hand; and clearly, computer science (and in particular artificial intelligence, or AI, where the logic of knowledge first appeared) explicitly borrowed from philosophical logic. But still the convergence is breathtaking, even in retrospect.

The other factor is the concrete distributed-systems interpretation of what otherwise are quite abstract models. A "possible world" is a global state of a system, a snapshot of it within an execution history. The global state of the system is made up of the local state of each of the processors. Two possible worlds are accessible from the perspective of a given processor if the process has the same local state in them. This obviously gives rise to the partition model of knowledge, and the S5 logic. Never mind that, on a closer look, S5 seems quite problematic as a model of knowledge (more on this later); it was remarkable to me that the "standard" logic of knowledge, which was proposed based on quite abstract reasoning, it fact arose naturally out of a very concrete and straightforward computational model. I will return to this point shortly.

2. What example(s) from your work, or work of others, illustrates the relevance of epistemic logic?

As a computer scientist I suppose it is incumbent on me to describe computer science applications, so let me do it, briefly. Two applications immediately come to mind. The first I already discussed – distributed systems. Computer scientists have always searched for good models with which to reason about distributed systems. The typical questions revolve around robustness to failures of various kinds, including failure by a node in the system to correctly follow the prescribed protocol, and failure of the network to transmit messages (or transmitting it with unbounded delay). Intuitively, people would make statements such as "Well, processor A doesn't know whether processer B received its message, but processor B knows that processor A doesn't know, and so..." Logics of knowledge proved an excellent way of making such reasoning precise, and specifically the formal construct of "common knowledge" (everyone knows, everyone knows that everyone knows, ...) completely characterized conditions necessary for coordinated action in the face of such system malfunctions.

Elsewhere (and earlier) in computer science the logic of knowledge was put into use in the context of AI planning. AI has a long tradition of planning research. In a typical scenario of so-called "classical" AI planning, the planner has some information about the state of the world, and some actions at its disposal, and must concoct a plan to change the world into some desired state by appropriate sequencing of the actions. Traditionally, each action has some preconditions (for example, you cannot grasp an object if you are already holding another object), which makes the problem computationally hard. Thirty years ago it was recognized that some preconditions are epistemic in nature; in order to open a safe you must know the combination of its lock. And, again, logics of knowledge proved just the ticket for modeling these preconditions. Later epistemic reasoning in planning included reasoning about epistemic post-conditions (after running a litmus test you know whether or not the substance is acidic), and coordinating the action of multiple robots with limited sensing capability.

3. What is the proper role of epistemic logic in relation to other disciplines, for instance mainstream epistemology, game theory, computer sciences or linguistics?

To me understanding the proper role of epistemic logic is a matter of understanding what I call the *rules of the game*, that is the methodological differences among the fields that epistemic logic contributes to. I will say more about this in my answer to Question 4.

4. Which topics and/or contributions should have had more attention in late 20th century epistemic logic?

I will list two technical topics, and one nontechnical. The technical topics include the connection between knowledge and belief, and iterated belief revision. The nontechnical item is a meta-discussion about the rules of the game. It is not that these topics did not receive attention; they did. But the discussion around them seemed to subside long before closure was reached, which makes it hard to build on those foundations as we try to apply them and/or extend the domain of discourse.

There are two common slogans concerning the connection between knowledge and belief, both reducing the former to the latter. The first is about knowledge being "justified, true belief".[1] Critique of this slogan (which I do not share) notwithstanding,

---

[1] For the record, I am aware of at least one suggestion in the opposite direction, defining belief as "defeasible knowledge", and an accompanying technical proposal, but I will not dwell on it.

I am not aware of a successful, generally accepted technical embodiment of this slogan (the problem is in capturing the notion of justification; as a side note, I believe that this avenue has been under-explored, and that the hasty rejection of this slogan was based on a an example, due to Gettier, which adopted a sense of justification that is inappropriate in this context).

The other slogan is of knowledge as "stable belief", or, more verbosely, "belief that is stable with respect to the truth". The idea is that one has beliefs, which over time get revised by new (correct) information that is learned. Those beliefs that are never forced out by new (correct) evidence are elevated to the status of knowledge. Several related technical proposals were made to capture this intuition; a typical one posited a total preorder on possible worlds as the accessibility relation, leading to a "standard" KD45 model of belief, but a model of knowledge which is weaker than S5. Specifically, the negative introspection property of knowledge is jettisoned, and replaced by weaker conditions, such as those captured by the S4.3 logic (with S4.3 lying in-between S4 and S5).

These latter proposals seem to me quite compelling, and it's puzzling to me why the discussion around them never picked up. Even before these proposals, S5 had been critiqued as a model of knowledge, and it was proposed that somewhat weaker logics be considered, specifically those lying in the neighborhood of S4.2 and S4.3. The fact that an independently motivated model happens to produce independently argued-for properties of knowledge is already striking. But there are additional attractive properties of these models. Beside inducing appealing properties on both knowledge and belief, they induce quite natural connections between them (knowledge being stronger than belief, and various natural hybrid introspective properties involving both knowledge and belief; for example, if you know something you believe that you do). And furthermore, the total-preorder structure had been considered earlier in the context of belief revision and nonmonotonic logics. So it is remarkable that a model with all these attractive properties and correlation with other independent proposals did not take hold more strongly than it did, especially since (to my knowledge) no alternative connection between knowledge and belief was proposed. It seems to me this reflects fatigue and apathy on the part of the research community more than disagreement.

Belief revision has certainly received tremendous attention. In this attempt to capture the dynamics of beliefs, researchers have investigated normative theories of revising a logical theory with a newly arrived piece of evidence. The interesting part of the theory concerns the case in which the new evidence is inconsistent with the original theory. The dominant (though not uncontested) theory in this area consists of the AGM axioms, which, as is well known, are closely related to the total-preorder structure discussed above (or, equivalently, Groves' original "system of spheres"). The limitations of the AGM theory become apparent when one tries to iterate the revision operator. The AGM axioms have nothing to say about it, and it is not obvious how to extend the model theoretic analysis. Briefly, the model theoretic account of AGM takes as input two arguments, a belief state (essentially, a total preorder on worlds) and a belief (a set of worlds). For the static theory, it is enough that the operator produce another set of worlds; that is enough to define the new beliefs. But if one wants to iterate the operation, the operator must produce an entire new belief state. There has been a continuous strand of literature on this topic, though not nearly as voluminous as in the 1980s. Opinions diverge on the state-of-the-art here. Some (mostly from AI) have argued that we have reached a good understanding of iterated belief revision. Others (including a well known philosophical logician) have argued that we have not. I tend to side with the latter. In any event, in such situations the nay sayers have an unfair advantage; if some people claim there is universal agreement, and others claim there is not, then there is not.

Finally, a word about the rules of the game, and about methodological differences among fields. Let me focus on philosophy and computer science. There is in certain philosophical circles an established tradition, when considering formal theories of everyday concepts, of relying on particularly instructive examples as litmus tests. The "morning star – evening star" example catalyzed discussion of cross-world identity in first-order modal logic. Closer to home, the example of believing that you will win the lottery and coincidentally later actually winning it served to disqualify the definition of knowledge as true belief. A similar example purported to disqualify defining knowledge as justified true belief (though, as I mentioned earlier, I don't think it did).

Such "intuition pumps" can be highly instructive, but the question is what role they play. In the above-mentioned philosophical circles they tend to serve as necessary but insufficient conditions for a theory. They are necessary in the sense that each of them is considered sufficient grounds for disqualifying a theory (namely, a theory which does not treat the example in an intuitively satisfactory manner). And they are insufficient since new examples can always be conjured up, subjecting the theory to ever-increasing demands.

This is understandable from the standpoint of philosophy, to the extent that it attempts to capture a complex, natural notion (such as knowledge) it its full glory. But this necessary-but-insufficient interpretation means that the process of formalization is a never-ending one, since new requirements may surface at any moment.

There is an alternative, and logics of knowledge provide an example. The S5 logic of captures well certain aspects of knowledge in idealized form, but the terms "certain" and "idealized" are important here. The logic has nothing to say about belief (as opposed to knowledge), nor about the dynamic aspects of knowledge (how it changes over time). Furthermore, even with regard to the static aspects of knowledge, it is not hard to come up with everyday counterexamples to each of its axioms. And yet, as discussed, the logic proves useful to reason about certain aspects of distributed systems, and the mismatch between the properties of the modal operator K and the everyday word "know" does not get in the way, within these confines. All this changes as one switches the context. For example, if one wishes to consider cryptographic protocols, the K axiom (no relation to the modal operator K) – which is valid in any normal modal logic, and here represents logical omniscience – is blatantly inappropriate.

The upshot of all this is the following criterion for a formal theory of natural concepts: One should be explicit about the intended use of the theory, and within the scope of this intended use one should require that everyday intuition about the natural concepts be a useful guide in thinking about their formal counterparts. A concrete interpretation of the above principle is what elsewhere I have called the "artifactual" perspective. Artifactual theories attempt to shed light on the operation of a specific artifact, and use the natural notion almost as a mere visual aid. This is precisely the case in knowledge and distributed systems. Does this make the artifactual perspective uninteresting from the philosophical point of view? I think not, though of course that is for philosophers to decide. In any event, it is a useful from the perspective computer science, which brings me to my last observation: While we have an artifactual perspective on the standard (if somewhat discredited) "standard" logic of knowledge, we do not have such a perspective on belief. This is another direction under-explored in the late 20th century.

5. What are the most important open problems in epistemic logic and what are the prospects for progress?

The previous section laid out several issues, including two technical ones, that remain unsettled, and that are important in my view. Naturally they constitute part of the answer here. But I will focus on a different direction, a vast open area in which many fascinating and important open problems lie. I am referring to the general program to develop logics of rational agency. These logics attempt to capture various facets of human mental state, and posit normative relationships among them (or "rational balance", to use Nilsson's term). Knowledge and belief belong to one category of metal state, which might be called "informational attitudes", capturing agents' assessment of whether this or that fact holds true. But informational attitudes constitute just one facet of mental state. In contrast, "motivational attitudes" capture something about the agents' preference structure, and "action attitudes" capture his inclination towards taking certain actions. In a typical theory, the action attitudes mediate between the informational and motivational attitudes; the agent's choice of action is dictates by his wants and beliefs. Into these two broad camps fall notions such as desire, goal, intention and plan. The literature on such attitudes is in comparison quite thin, and herein lies the opportunity. There is a reason why progress in this area has been more limited. When one thinks about preference in isolation of other factors, things are relatively easy. There are certainly some interesting challenges – for example, capturing "ceteris paribus" conditions in preference (I prefer wealth to poverty all other things being equal, but I prefer being healthy and poor to being sick and rich). But things become truly involved when one considers the interaction between the various types of attitude. For example, the dynamics of belief and preference are in general intertwined, with changes in beliefs leading to change in preference (and possibly vice versa). But more complex interactions are common. In a typical situation, one "intends" to take an action in service of a "goal" which was given rise to by certain "desires", and conditional on some " beliefs" (for example, intending to drive your car to the city is motivated by your belief that there is a good show there that evening, and is inconsistent with believing that the car is not in working condition). Beside the interaction between the different attitudes (belief, intention, goal, desire), the discussion involves even more basic aspects of agency, such as action and ability. This is a complicated picture, and so it's not surprising that progress on this has been relatively slow. But slow progress does not mean no progress. The literature on so-called BDI theories (a tongue-in-cheek reference to belief, desire and intention) contains some inspiring and intriguing ideas and provides several starting points. I believe there is much low-hanging, and tasty, fruit in this area.

23
John Sowa

Croton on Hudson, New York, USA

---

1. Why were you initially drawn to epistemic logic?

My work on epistemic logic developed from my research in artificial intelligence. My formal education in mathematics and my interests in philosophy and linguistics led me to Richard Montague's synthesis of a formal grammar with an intensional semantics based on Kripke's possible worlds. Around the same time, however, my readings in psycholinguistics showed that young children learn to use modal expressions with far greater ease and flexibility than graduate students learn Montague's logic. Following are some sentences spoken by a child named Laura who was less than three years old (Limber 1973):

> Here's a seat. It must be mine if it's a little one.
>
> I want this doll because she's big.
>
> When I was a little girl I could go "geek-geek" like that. But now I can go "this is a chair."

In these sentences, Laura correctly expressed possibility, necessity, tenses, indexicals, conditionals, causality, quotations, and metalanguage about her own language at different stages of life. She already had a fluent command of a much larger "fragment of English" than Montague had formalized.

The syntactic theories by Chomsky and his followers were far more detailed than Montague's, but their work on semantics and pragmatics was rudimentary. Some computational linguists had developed a better synthesis of syntax, semantics, and pragmatics, but none of their programs could interpret, generate, or learn language with the ease and flexibility of a child. The semantic networks of artificial intelligence provided a flexible computational framework with a smooth mapping to and from natural languages.

In 1976, I published my first paper on conceptual graphs as a computable notation for natural language semantics, but I hoped to find a simpler and more natural representation for the logical operators and rules of inference.

In 1978, I discovered Peirce's existential graphs. Peirce had developed three radically different notations for logic: the relational algebra, the algebraic notation for predicate calculus, and the existential graphs, which he developed in detail during the last two decades of his life. Although his original existential graphs were limited to first-order logic, Peirce added further innovations for representing modality, metalanguage, and higher-order logic (Roberts 1973). Other logicians have since developed these topics in greater detail, but often in isolation from broader issues of science and philosophy. Peirce, however, sought to integrate every aspect of semiotics by developing his graphs as a "system for diagrammatizing intellectual cognition" (MS 292:41). My studies in AI, linguistics, and philosophy convinced me that such a synthesis is a prerequisite for understanding human intelligence or developing an adequate computer simulation.

2. What example(s) from your work, or work of others, illustrates the relevance of epistemic logic?

Epistemic logic belongs to the tradition from Aristotle to Peirce of using logic to analyze science, language, thought, and reasoning. Unlike Frege and Russell, who focused on the foundations of mathematics, Peirce integrated logic with semiotics and applied it to every area of science, philosophy, and language. To represent modal contexts in existential graphs, Peirce experimented with colors and dotted lines. Frege insisted on a single domain of quantification called everything, but Peirce used tinctures to distinguish different universes:

> The nature of the universe or universes of discourse (for several may be referred to in a single assertion in the rather unusual cases in which such precision is required) is denoted either by using modifications of the heraldic tinctures, marked in something like the usual manner in pale ink upon the surface, or by scribing the graphs in colored inks. (MS 670:18)

Peirce considered three universes: actualities, possibilities, and the necessitated. He subdivided each universe in four ways to define 12 modes. In the universe of possibilities, for example, he distinguished objective possibility (an alethic mode), subjective possibility (epistemic), social possibility (deontic), and an interrogative mode, which corresponds to scientific inquiry by hypothesis and experiment. For the necessitated, he called the four subdivisions the rationally necessitated, the compelled, the commanded, and the determined. Most of his writings on these topics were unpublished, and he changed his terminology from one manuscript to the next. Peirce admitted that a complete analysis and classification would be "a labor for generations of analysts, not for one" (MS 478:165).

During the 20th century, Peirce's broad approach influenced logicians and philosophers who developed theories of epistemology. The first was Clarence Irving Lewis, whose teachers at Harvard included the pragmatists William James and Josiah Royce. After earning his PhD in 1910, Lewis lectured at Berkeley, where wrote his Survey of Symbolic Logic in 1918. In the historical introduction, he devoted 21 pages to Boole, 28 pages to Peirce, and 11 pages to Whitehead and Russell. Most of the expository material covered classical first-order logic, but Lewis devoted 49 pages to his own research on strict implication, which eventually led to his pioneering work on modal logic. In 1920, he joined the Harvard faculty, where he spent two years studying the manuscripts donated by Peirce's widow. After those studies, Lewis began to call himself a conceptual pragmatist, and he continued to write and lecture on modal logic, epistemology, and related topics for the rest of his life (Murphey 2005).

Other logicians with interests in epistemology and modal logic were also strongly influenced by Peirce. Arthur Prior quoted Peirce extensively in his writings on propositions (1976) and temporal logic (1977). In a 16-page critique of Wittgenstein's Tractatus, Frank Ramsey noted that Peirce's type-token distinction could clarify an ambiguity in Wittgenstein's analysis of propositions. In his later discussions with Wittgenstein, Ramsey recommended Peirce's writings (Nubiola 1996). Wittgenstein apparently accepted that advice because he recommended some papers by Peirce (1923) in a letter to his sister. Jaakko Hintikka has also written about both Peirce and Wittgenstein. Like Peirce, these philosophers applied logic to a broader range of issues than just the foundations of mathematics.

3. What is the proper role of epistemic logic in relation to other disciplines, for instance mainstream epistemology, game theory, computer sciences or linguistics?

As a version of modal logic for reasoning about knowledge and belief, epistemic logic is intimately related to other modal logics and to the methods for acquiring, learning, and reasoning about knowledge. Every issue related to those topics is directly or indirectly relevant to epistemic logic; e.g., the criticisms by Dana Scott (1973):

> Formal methods should only be applied when the subject is ready for them, when conceptual clarification is sufficiently advanced... No modal logician really knows what he is talking about in the same sense that we know what mathematical entities are. This is not to say that the work to date in modal logic is all bad or wrong, but I feel that insufficient consideration has been given to questioning appropriateness of results... it is all too tempting to refine methods well beyond the level of applicability.

C. I. Lewis (1960) made a similar observation:

> It is so easy... to get impressive "results" by replacing the vaguer concepts which convey real meaning by virtue of common usage by pseudo precise concepts which are manipulable by "exact" methods — the trouble being that nobody any longer knows whether anything actual or of practical import is being discussed.

Quine (1972) noted that Kripke models can be used to prove that modal axioms are consistent, but they don't explain the intended meaning of those axioms:

> The notion of possible world did indeed contribute to the semantics of modal logic, and it behooves us to recognize the nature of its contribution: it led to Kripke's precocious and significant theory of models of modal logic. Models afford consistency proofs; also they have heuristic value; but they do not constitute explication. Models, however clear they be in themselves, may leave us at a loss for the primary, intended interpretation.

Kripke's major achievement was to demonstrate the consistency of modal logic and the axioms by C. I. Lewis in terms of set theory. But he assumed two primitives: a set of undefined entities called possible worlds and an undefined relation R among worlds. If R(w1,w2) happens to be true, then the world w2 is said to be accessible from the world w1. Kripke made some informal remarks to justify those primitives, but Scott and Quine would agree with Lewis: "nobody any longer knows whether anything actual or of practical import is being discussed."

A major limitation of most modal axioms and Kripke models is that they only relate a single modality. They permit dubious combinations such as possibly necessary, but they can't represent the sentence I know that I'm never obligated to do anything impossible, which combines epistemic, temporal, deontic, and alethic modes. Scott (1970) considered the limitation to just one modality "one of the biggest mistakes of all in modal logic":

The only way to have any philosophically significant results in deontic or epistemic logic is to combine these operators with: Tense operators (otherwise how can you formulate principles of change?); the logical operators (otherwise how can you compare the relative with the absolute?); the operators like historical or physical necessity (otherwise how can you relate the agent to his environment?); and so on and so on.

To represent and reason about multiple modalities and their interactions, some logicians have added multiple accessibility relations to the Kripke models. Cohen and Levesque (1990), for example, introduced a belief accessibility relation for reasoning about knowledge and belief and a goal accessibility relations for desires and intentions. To justify the formalism, they related their axioms to the independently developed theory of beliefs, desires, and intentions by Bratman (1987). That justification is important, but Laura at age three could already use modal verbs to express her beliefs, desires, and intentions. The linguistic mechanisms that support her language are likely to be simpler than Kripke models with multiple accessibility relations.

Modern logicians have gone far beyond Peirce in developing axioms for various modalities, but they neglected his goal of classifying and relating the modalities in a systematic framework. Although Peirce never completed the framework, his four-way subdivision is still worth exploring:

1. Objective reality (actual, possible, or necessary) is independent of what any observer thinks. These are the alethic

modes.

2. Each individual's subjective experience is a personal actuality in which beliefs develop, are verified, and become knowledge. These are the epistemic modes.

3. Each social group, ranging from a nuclear family to nation states, develops a consensus of habits, conventions, norms, regulations, and laws. These define the deontic modes.

4. Peirce regarded science as a collaboration among inquirers working on common interests and problems over a time span of millennia. At any point in time, their consensus about observations, hypotheses, and laws is a good, but fallible approximation to the objective modes.

This classification can be applied to any individual or social group at any stage of development. It can be used to analyze the way a child learns and uses modal terms. It can also be used to study social norms, legal systems, or scientific practice and methodology.

4. Which topics and/or contributions should have had more attention in late 20th century epistemic logic?

I have found the most fertile sources of new ideas for further research in historical writings that have been forgotten or in unpopular approaches that have been ignored. When I was doing the early research on conceptual graphs, I came across a collection of papers on "alternative semantics" edited by Hugues Leblanc (1973). Some of them, such as the one by Scott, contained stimulating advice and criticism. The two I used to define the semantics of CGs were by Michael Dunn and Jaakko Hintikka.

Dunn (1973) showed that the set of possible worlds in a Kripke model could be replaced by a pair of sets (M, L). For each world w in Kripke's sense, M is a Hintikka-style model set of all the facts that are true about w. Dunn's innovation is to designate L as the set of laws of w, which are necessarily true of w. For any Kripke model, there is always a Dunn model with an isomorphism of each world w to a pair (M, L) of the facts and laws of w. This isomorphism guarantees that any theorems derived in terms of Kripke models are also true of Dunn models. For that reason, most logicians have ignored Dunn models.

What makes Dunn models interesting is that the accessibility relation is no longer primitive. It can be derived from the choice of laws. Dunn defined accessibility from any world w_i to a world w_j to mean that the laws L_i of the first world are a subset of the facts M_j of the second world:

$$R(w_i, w_j) \equiv L_i \subset M_j.$$

According to this definition, the laws of the first world w_i are true in the second world w_j, but they may be demoted from the status of laws to merely contingent facts. Dunn's definition of accessibility holds for normal models, in which every law is also a fact, and for non-normal models, in which some laws might not be facts. In deontic logic, for example, the laws are obligatory, but sinners might violate them. For such models, the accessibility relation R is not reflexive because a non-normal world is not accessible from itself. This point is one more example of the nonintuitive nature of Kripke models: it seems odd to say that any world could be inaccessible from itself, especially the real world because it happens to have sinners. But there is nothing odd or counterintuitive about saying that some laws might be violated.

Although Dunn models and Kripke models are isomorphic when only one modality is considered, Dunn models are much easier to generalize for multimodal reasoning (Sowa 2003, 2006). With Kripke models, a new accessibility relation must be added as an undefined primitive for each type of modality. With Dunn models, each law can be tagged with a metalevel marker that indicates its source as a law of physics, chemical engineering, the United States, some local town, or some business policy. If no context is indicated, a general modal verb such as must would indicate necessity as determined by the conjunction of all applicable laws. For a particular context, a subscript could indicate that some subset of laws L_i determines the necessity operator $\Box_i$ and the possibility operator $\Diamond_i$. With this approach, no special primitives or assumptions are needed. Metalevel reasoning can be used to select the applicable laws for any particular problem, and the object-level reasoning would follow the usual first-order rules of inference.

The paper by Hintikka (1973) introduced two techniques that I adopted for defining the semantics of conceptual graphs and using them in processing natural language: surface models and game theoretical semantics. Hintikka's surface models anticipated the dynamic semantics of Groenendijk and Stokhof (1991). In both methods, the process of interpreting a sentence constructs a partial model of a world in which the sentence happens to be true. Hintikka's tree representation for surface models is easily adapted to graphs. His game-theoretical semantics happens to be closely related to Peirce's method of endoporeutic (outside-in evaluation) for evaluating the truth of an existential graph in terms of a model that could also be represented as a graph. The combination of these techniques provides a dynamic method for interpreting natural language text or dialog in graph models and evaluating the truth of logical graphs in terms of such models. It is also a realistic psycholinguistic hypothesis about the way people understand language, and it supports Peirce's claim that graphs can capture a "moving picture of the action of the mind in thought" (Pietarinen 2005).

5. What are the most important open problems in epistemic logic and what are the prospects for progress?

The most important task is to integrate the principles and methods discussed in the previous four answers: the relevance of epistemic logic to learning, using, and communicating knowledge; the psycholinguistic issues raised by child language and learning; the computational efficiency and flexibility of computer implementations; the empirical justification of the abstract formalisms; the classification of all modalities including the epistemic; the dynamic changes in knowledge and belief during a dialog; and the use of these principles in the design of intelligent agents that can communicate with people in ordinary language.

As Peirce observed, this is a task for "generations of analysts." But several generations have already passed, and we can use his guidelines to evaluate the progress. The first is Peirce's principle of pragmatism, which C. I. Lewis adopted: "The elements of every concept enter into logical thought at the gate of perception and make their exit at the gate of purposive action; and whatever cannot show its passports at both those two gates is to be arrested as unauthorized by reason" (CP 5.212). Epistemic logic, by itself, isn't directly connected to either gate. Lewis made some connections by alternating his work on abstract logic with his writings on epistemology. But a more complete solution must situate epistemic logic within a framework that includes both gates and all the interconnections.

No deductive logic can link all the steps from perception to action. Peirce emphasized the three methods of reasoning: induction, abduction, and deduction. The first two are necessary to derive the premises, from which deduction can derive the consequences. But those consequences must be tested against further observations, which lead to another round of induction, abduction, and deduction. Instead of an epistemic logic, the result is an epistemic cycle:

1. Induction. Perception supplies the observations from which induction derives generalizations. But memory to store previous observations and analogy to find similar observations are also required to gather the data for induction.

2. Abduction. For any particular purpose, abduction is required to gather and combine relevant observations and generalizations in novel hypotheses or to revise and refine previous hypotheses. The results are the premises used by deduction.

3. Deduction. Given a theory, deduction can derive the consequences, answer questions, or make predictions. The results may be stored in memory as a basis for further reasoning, but testing is necessary to ensure their accuracy and relevance.

4. Action. The results of deduction are a guide to actions upon the world. Those actions may lead to further observations, which may confirm, contradict, or refine the predictions. Then the cycle continues with a new round of inductions, abductions, deductions, and testing.

Note that deduction covers only 25% of the cycle from perception to action. The most important problem for epistemic logic is to explore the entire cycle and all its connections.

References

Bratman, Michael E. (1987) Intentions, Plans, and Practical Reason, Harvard University Press, Cambridge, MA.

Cohen, Philip R., & Hector J. Levesque (1990) Intention is choice with commitment, Artificial Intelligence 42:3, 213-261.

Dunn, J. Michael (1973) A truth value semantics for modal logic, in Leblanc (1973) pp. 87-100.

Groenendijk, Jeroen, & Martin Stokhof (1991) Dynamic predicate logic, Linguistics and Philosophy 14:1, 39-100.

Hintikka, Jaakko (1973) Surface semantics: definition and its motivation, in Leblanc (1973) pp. 128-147.

Leblanc, Hugues, ed. (1973) Truth, Syntax and Modality, North-Holland, Amsterdam.

Lewis, Clarence Irving (1918) Survey of Symbolic Logic, University of California Press, Berkeley.

Lewis, Clarence Irving (1960) Personal letter, quoted by Hao Wang (1986) Beyond Analytic Philosophy: Doing Justice to What We Know, MIT Press, Cambridge, MA, p. 116.

Limber, John (1973) The genesis of complex sentences, in T. Moore, ed., Cognitive Development and the Acquisition of Language, Academic Press, New York, 169-186.

Murphey, Murray G. (2005) C. I. Lewis: The Last Great Pragmatist, State University of New York, Albany.

Nubiola, Jaime (1996) Scholarship on the relations between Ludwig Wittgenstein and Charles S. Peirce, in I. Angelelli & M. Cerezo, eds., Proceedings of the III Symposium on the History of Logic, Gruyter, Berlin.

Peirce, Charles Sanders (1923) Chance, Love and Logic: Philosophical Essays, edited by Morris R. Cohen, Harcourt, Brace, & Co., New York.

Peirce, Charles Sanders (CP) Collected Papers of C. S. Peirce, ed. by C. Hartshorne, P. Weiss, & A. Burks, 8 vols., Harvard University Press, Cambridge, MA, 1931-1958.

Peirce, Charles Sanders (EP) The Essential Peirce, ed. by N. Houser, C. Kloesel, and members of the Peirce Edition Project, 2 vols., Indiana University Press, Bloomington, 1991-1998.

Pietarinen, Ahti-Veikko (2005) Peirce's magic lantern of logic: Moving pictures of thought, Transactions of the Charles S. Peirce Society.

Prior, Arthur N. (1976) The Doctrine of Propositions and Terms, edited by P. T. Geach & A. J. P. Kenny, University of Massachusetts Press, Amherst.

Prior, Arthur N., & Kit Fine (1977) Worlds, Times, and Selves, University of Massachusetts Press, Amherst.

Roberts, Don D. (1973) The Existential Graphs of Charles S. Peirce, Mouton, The Hague.

Scott, Dana S. (1970) Advice on modal logic, in K. Lambert, ed., Philosophical Problems in Logic, D. Reidel, Dordrecht, pp. 143-173.

Scott, Dana S. (1973) Background to formalization, in Leblanc (1973) pp. 244-273.

Sowa, John F. (1976) Conceptual graphs for a database interface, IBM Journal of Research and Development 20:4, 336-357.

Sowa, John F. (1984) Conceptual Structures: Information Processing in Mind and Machine, Addison-Wesley, Reading, MA.

Sowa, John F. (2000) Knowledge Representation: Logical, Philosophical, and Computational Foundations, Brooks/Cole Publishing Co., Pacific Grove, CA.

Sowa, John F. (2003) Laws, facts, and contexts: Foundations for multimodal reasoning, in Knowledge Contributors, edited by V. F. Hendricks, K. F. Jørgensen, and S. A. Pedersen, Kluwer, Dordrecht, pp. 145-184.

Sowa, John F. (2006) Worlds, models, and descriptions, Studia Logica, Special Issue Ways of Worlds II, 84:2, 323-360.

24

Robert Stalnaker

Laurance S. Rockefeller Professor of Philosophy
Department of Philosophy and Lingusitics, MIT, USA

---

1. Why were you initially drawn to epistemic logic?

Jaakko Hintikka's book, *Knowledge and Belief*, was of course the starting point, and it was one of the things that helped me see the fruitfulness of the possible worlds representation of the contents of both speech acts and mental states. But it was my involvement in the first TARK conference, in 1986 (originally, "Theoretical Aspects of Reasoning about Knowledge"; later, changed to "Theoretical Aspects of Rationality and Knowledge") that got me to appreciate the clarifying power of epistemic logic, and its semantics, and its applications outside of philosophy. I served on the program committee for the first incarnation of this continuing conference, and most of the papers and abstracts I read were in theoretical computer science, an area I had previously known nothing about. These theorists found it useful to think of distributed systems as communities of knowers, exchanging information, including information about what information was available to other processors in the system. Epistemic logic and its semantics, with multiple knowers (and so multiple epistemic accessibility relations) provided a precise representation, at the right level of abstraction, of the flow of information through the system, and of the character of various protocols governing such systems. The "possible worlds" were the states of the system as a whole, or sometimes the possible histories of the system as a whole, that were compatible with the transition rules governing the system. The general lesson I drew from this work was that it was useful for epistemology to think of communities of knowers, exchanging information and interacting with the world, as (analogous to) distributed computer systems. The models developed in these theories were abstract and idealized enough to allow for precision, but they had a richer structure than the more abstract models of pure epistemic logic (that treated possible worlds as primitive points), a structure that made it easier to connect the semantic framework of epistemic logic with temporal and causal notions.

Joe Halpern, who was the moving force behind the organization of the TARK meetings, deserves a lot of credit, not only for his own work in this area, but for seeing before others the mutual relevance of work in philosophy, theoretical computer science, and economics, and for facilitating the interaction of theorists who had previously been ignorant of each other's work. This led to what continues to be a thriving interdisciplinary enterprise, in which epistemic logic and its semantics play an important role.

2. What example(s) from your work, or work of others, illustrates the relevance of epistemic logic?

One of the most interesting applications of epistemic logic and semantics is to the foundations of game theory, and this the area of most of my own work that uses epistemic logic. Traditional game theory uses a representation of a game that specified utility values for the different players for all the alternative outcomes of a game, but had no explicit representation of the beliefs of the players about what the other players will do. Instead, the theory made only an informal intuitive assumption that the players had full knowledge of the structure of the game, and that they took it as common knowledge that all players would act rationally. But there was no formal representation either of knowledge or of rationality. The newer approach, which Robert Aumann labeled "interactive epistemology", was to model the playing of games with a representation of the knowledge, beliefs, and degrees of belief of the players. One can then see game theory as an application of ordinary Bayesian individual decision theory, with rationality defined as maximizing expected utility. The game theoretic assumption that the players have common knowledge (or belief) of the rationality of all players can then be represented formally as a constraint on the admissible beliefs, and so the admissible actions, of the players. Given the semantics of epistemic logic, it is straightforward to represent common knowledge or belief in terms of individual knowledge or belief, but the application of this framework to game theory raised conceptual and philosophical questions about the relation between knowledge and belief, issues that have consequences for the evaluation of some of the strategic arguments, such as backward induction arguments, that have long been familiar in the game theory literature. John Harsanyi's work on type spaces and Robert Aumann's early paper on agreeing to disagree were important influences in the development of the interactive epistemology approach. An excellent survey paper about the approach is the following: P. Battigalli and G. Bonanno, "Recent results on belief, knowledge and the epistemic foundations of game theory," *Research in Economics*, **53** (1999), 149-225. I have written several papers in this area, including "Knowledge, Belief and Counterfactual Reasoning in Games," *Economics and Philosophy*, **12** (1996), 133-162, and "On the Evaluation of Solution Concepts," in *Epistemic Logic and the Theory of Games and Decisions*, ed. by M. O. L. Bacharach, L.-A. Gérard-Varet, P. Mongin and H. S. Shin. Kluwer Academic Publisher, 1997, 345-64.

I also have a paper, ("On Logics of Knowledge and Belief, *Philosophical Studies* **120** (2006), 169-199) that tries to connect structural assumptions about the epistemic accessibility relation for interpreting epistemic logic with substance epistemological assumptions about knowledge. The general aim is show the relevance of the formal work to mainstream epistemology.

3. What is the proper role of epistemic logic in relation to other disciplines, for instance mainstream epistemology, game theory, computer sciences or linguistics?

As noted above, epistemic logic has applications both within epistemology and outside of philosophy, and I think they are connected. Both epistemologists and game theorists are concerned to clarify the relation between knowledge and practical action, and the issues that come up in applying epistemic semantics to game theory connect with issues discussed by mainstream epistemologists. In linguistic semantics and pragmatics, the connections are more direct, since knowledge and other propositional attitudes such as speaker presupposition are part of the theoretical apparatus. Context is represented by a set of possible worlds that is defined by an informational state that has the structure of common knowledge or belief. Theoretical computer scientists are concerned with error detection and avoidance in computer systems; obviously, perfection is not achievable, and one aims to avoid total crashes when something goes wrong. The problems of saying what a system "knows" when something (but not everything) goes wrong has much in common with the epistemologist's problem of saying what counts as knowledge in a case where some (but not all) of our sources of information are corrupted. I think philosophers concerned with epistemology and researchers outside of philosophy who are applying epistemic concepts have things to learn from each other, and the common framework of epistemic logic and semantics facilitates that interaction.

4. Which topics and/or contributions should have had more attention in late 20th century epistemic logic?

A classic paper, later than Hintikka's early work, but earlier than the resurgent interest in epistemic logic, that I think has not received the attention it deserves is W. Lenzen's "Recent work in epistemic logic," *Acta Philosophica Fennica* **30**(1978). (I believe Lenzen's results were originally published earlier in German) This paper reports some important results of his about the relation between knowledge and belief that have sometimes been ignored.

5. What are the most important open problems in epistemic logic and what are the prospects for progress?

Epistemic logics are, for the most part, normal modal logics, which means that they contain a necessitation rule (if ϕ is a theorem, then $K\phi$ is a theorem) and a distribution principle

$$K(\phi \to \psi) \to (K\phi \to K\psi).$$

These principles saddle epistemic semantics with the problem of logical omniscience, since in the idealized theory, it will follow that knowers know all of the logical consequences of their knowledge. This is a very general problem in formal epistemology, applying also to the use of probability theory to represent degrees of belief and epistemic confirmation relations, and I think it is one of the most important outstanding general problems for epistemic logic. I think most attempts to weaken the logic to avoid the consequence are unsuccessful, since they tend to be ad hoc, and to be incompatible with any plausible semantic interpretation. The problem is a conceptual rather than a technical one, and I think there is still work to be done in saying exactly what the problem is. That is, we need to get clearer, in general, about the relation between formal models and the phenomena that they aim to model. It is interesting, and puzzling, that theories that make such egregiously unrealistic assumptions can nevertheless seem to be fruitful and clarifying. That does not mean we can ignore the problem, but I think it does give us reason to carry on, with the hope that a solution, or at least a clearer view of the problem, will emerge as the framework is developed and applied.

My own discussions of this problem (most specifically, "Logical Omnsicience I" and "Logical Omniscience II", both in Stalnaker, *Context and Content*, OUP, 1999) are only attempts to say what the problem is, and to argue that it is conceptual problem we all face, and not one that is subject to a technical fix.

I think the main promise of further insight from the application of epistemic logic will come from multi-modal theories that combine epistemic with causal and temporal concepts, and concepts concerning action. One for the virtues of the general modal semantic framework is that it provides a common structure in which concepts from these various domains can be combined.

Timothy Williamson

Wykeham Professor of Logic
University of Oxford, UK

---

1. Why were you initially drawn to epistemic logic?

My first publication was in epistemic logic, broadly conceived ('Intuitionism Disproved?', *Analysis*, 1982). It was about Frederic Fitch's so-called paradox of knowability, an argument that the apparently moderate verificationist thesis that all truths are knowable entails the absurdly extreme verificationist thesis that all truths are known. Fitch thanked an anonymous referee for *The Journal of Symbolic Logic* for the argument; very recently it has been discovered that the anonymous referee was Alonzo Church. I pointed out that the argument does not go through in intuitionistic logic, which was significant because the best developed case for moderate verificationism consisted in Michael Dummett's considerations in favour of anti-realism, which he also took to motivate the replacement of classical logic by intuitionistic logic. Although I had no sympathy for anti-realism, I thought that dismissing it on the basis of the Church-Fitch argument was too quick. Technically I was right, although these days I have less patience for the eccentricities of anti-realism. The Church-Fitch argument is formulated within a bimodal logic with operators for both knowledge and possibility; it uses formal principles about knowledge (what is known is true; a known conjunction has known conjuncts). To explore different interpretations of the argument I added a subscript for time to the knowledge operator, combined with quantification over times, so in those respects I was using a much richer language than that of pure epistemic logic.

Later, I started to use the standard language of monomodal logic while interpreting the modal operators epistemically in order to explore the consequences of other forms of verificationism, aiming for a *reductio ad absurdum*. My first publication of that kind was 'Assertion, Denial and Some Cancellation Rules in Modal Logic' (*Journal of Philosophical Logic*, 1988). It was prompted by a conversation with Lloyd Humberstone and subsequent correspondence with him (before the days of email!). I considered a form of verificationism according to which the truth-conditions of a sentence are determined jointly by its verification-conditions and its falsification-conditions, which I expressed as a rule of proof in monomodal logic, interpreting the necessity operator in terms of verification. I showed that if verification obeys the principles of the modal system S4, as verificationists typically assume, then that rule of proof leads to wildly implausible consequences.

Both articles led me to a sequence of further ones, the former on the Church-Fitch argument, the latter on rules of proof in modal logic. As these examples suggest, my interest in epistemic logic has been driven primarily by philosophical concerns. In that respect it had more in common with the early period of epistemic logic, in which Jaakko Hintkka's *Knowledge and Belief* (1962) was the key text, than with many of the developments over the past twenty years — comparatively few of the contributors to this volume would be regarded as primarily philosophers. Of course, my aim was to show how those philosophical questions were related to well-defined technical problems, and to make progress on the former by solving the latter.

As the examples also suggest, the philosophical concern that fired my original interest in epistemic logic was the debate between realism and anti-realism. From an Oxford perspective in the period 1973-80, when I was a philosophy student there, it seemed to be *the* central issue in philosophy. Michael Dummett, the leading proponent of anti-realism, was my supervisor for the final year of my doctoral studies. My dissertation was not on epistemic logic, but on the concept of approximation to the truth, Karl Popper's idea that scientific theories might approximate more and more closely to the truth even if they never quite got there, which provided one way of defending a realist philosophy of science. My undergraduate degree, also at Oxford, was in mathematics and philosophy, with a strong logic component, so I had the technical background to prove results in epistemic logic when needed.

Over the past twenty years, my philosophical interests have moved in the direction of epistemology. Those changes provided new scope for applying epistemic logic. I worked on vagueness, in part because it looked as though if anti-realism were right about anything, it would be right about vagueness, so that by showing it to be wrong about vagueness one could show it to be wrong about everything. In *Vagueness* (1994), I defended an epistemic theory of vagueness, on which vague terms have sharp boundaries whose location is unknowable. Vagueness consists in borderline cases, where the term neither clearly applies nor clearly fails to apply, but 'clearly' there is to be understood epistemically. Thus the logic of the 'clearly' operator is an epistemic logic, which I have explored in various publications. The further development of the epistemology behind *Vagueness* led to *Knowledge and its Limits* (2000). I used epistemic logic there to construct formal models of my approach to knowledge. In particular, it lent itself to a theory of probabilities on evidence within the framework of epistemic logic, a theory that avoids the epistemological naivety of standard Bayesian accounts.

Another stimulus to work on epistemic logic came in the early 1990s from the theoretical economist Hyun Song Shin, when we were both fellows of University College Oxford. Through him I learned much more about applications of epistemic logic outside philosophy. We collaborated on two joint papers. 'Representing the Knowledge of Turing Machines' (*Theory and Decision*, 1994) was written because he had been thinking about how to do epistemic logic for agents whose computational power does not exceed that of a Turing machine. A tempting argument seemed to show that the appropriate logic would be provability logic (in the sense of George Boolos and others), which does not even permit the axiom that what is known is true. Meanwhile, I had been thinking about what was wrong with the Lucas-Penrose argument, using Gödel's second incompleteness theorem, which supposedly shows that human agents do exceed Turing machines in computational power. Putting the two lines of thought together, we showed why the restriction to agents with the computational power of a Turing machine does not collapse epistemic logic into provability logic but does force some restrictions. For example, S5 — the most popular system of single-agent epistemic logic — is unsuitable for such agents under rather general conditions, but S4 is suitable, despite not being a subsystem of provability logic. The trick is to realize that which epistemic logic is valid for a given type of agent depends on how the knowledge operator is interpreted as representing the agent to itself when embedded in another occurrence of the operator. I have subsequently published various papers generalizing these ideas. The other paper that I co-authored with Hyun Shin is 'How Much Common Belief is Necessary for a Convention?' (*Games and Economic Behavior*, 1996). It arose from his realization that the models of epistemic logic I constructed for epistemological purposes in 'Inexact Knowledge' (*Mind*, 1992) can be used to define cognitive conditions under which simple rules can be stable in a community while complex rules cannot be.

Evidently, I have been drawn to epistemic logic at different times for somewhat different specific reasons. But in every case much of its general attraction has consisted in its capacity to give rigour and objectivity to the discussion of a philosophical question. Naturally, one must be careful not to lose sight of the philosophical question in one's search for rigour and objectivity!

2. What example(s) from your work, or work of others, illustrates the relevance of epistemic logic?

My answer to the previous question contains the beginnings of an answer to this one too. I will concentrate on some further examples. Shamelessly, I will take them all from my own work, since it represents an epistemological angle on epistemic logic that I regard as underrepresented in the field. Epistemic logic is relevant to lots of other subjects in lots of other important ways too, but I am sure that they will be well represented by other contributors to this volume.

In *Identity and Discrimination* (1990) I discussed phenomena such as the non-transitivity of indiscriminability. For instance, imagine a series of many colour samples from red to yellow arrayed in a circle like the spokes of a wheel and consider an observer looking at them by naked eye only, under fixed conditions. Each sample is indiscriminable in colour from its immediate predecessor and from its immediate successor in the series, but at the very same time the first member of the series is clearly discriminable in colour from the last member of the series (which it is placed next to). Therefore there are three samples x, y and z such that x is indiscriminable in colour from y, y is indiscriminable in colour from z and x is discriminable in colour from z. Such examples are crucial to discussion of the vagueness of terms such as 'red' and 'yellow'. I argued that discrimination is an epistemic phenomenon, not a behavioural one: to discriminate things in a given respect is to activate the knowledge (not just the belief!) that they are different in that respect. Correspondingly, indiscriminability is the epistemic possibility of sameness in the given respect. I therefore used epistemic logic combined with the logic of identity to predict the logical properties of indiscriminability, such as reflexivity, symmetry and non-transitivity, and to explore its logic more generally.

The problem of vagueness gains much of its depth from the phenomenon of higher-order vagueness: borderline cases of borderline cases, and so on. Its effect is that vagueness infects descriptions of vagueness too, in ways that most theories of vagueness are unable to handle. In 'On the Structure of Higher-Order Vagueness' (*Mind*, 1999) I used a framework of epistemic logic to define nth-order vagueness for each natural number n and to investigate the relations between different orders of vagueness and between higher-order vagueness in a complex sentence and higher-order vagueness in its constituents in a more systematic way than had previously been available.

Higher-order vagueness is closely related to failures of the widely accepted axiom variously known as 4, the KK principle and positive introspection, according to which if one knows something then one knows that one knows it. In 'Inexact Knowledge', *Vagueness* and *Knowledge and its Limits* I show that such failures are predictable from limitations on the agent's powers of discrimination in both perception and thought. Recently I have extended these results by using very natural epistemic models of perceptual discrimination to demonstrate the existence of cases in which one knows something even though the probability on one's evidence that one knows it is arbitrarily close to zero ('Improbable Knowing', to appear). Such cases help to explain the seductiveness of sceptical arguments: the temptation to deny that we know may be almost overwhelming even when we do know. More generally, they can be used to support an externalist approach to epistemology, on which even the most central facts about one's epistemic position may be inaccessible to one's knowledge.

A topic of increasing importance in epistemology is the role of deduction as a way of extending one's knowledge. It is hard to study within the framework of standard epistemic logic, since the latter validates the unrestricted closure principle of logical omniscience, according to which one automatically knows every logical consequence of what one knows. By contrast, the point of deduction as a way of extending one's knowledge is that closure is not automatic; one *extends* one's knowledge deductively only once one has competently performed the relevant deduction. We need a more fine-grained framework of epistemic logic within which to model such deductive extensions. I have been experimenting with a relation of epistemic counterparthood between formulas to do just that (*Probability and Danger*, the 2009 Amherst Lecture in Philosophy). The new semantics invalidates all non-trivial forms of logical omniscience while enabling one to define a notion of competence with respect to a connection between premises and conclusion which combines with knowledge of the premises to imply knowledge of the conclusion. The connection may be deductive or just a locally reliable material implication. This sort of deductive competence can be regarded as knowledgeability, a sort of generalization of knowledge, since its special case when the premise set is empty is simply knowledge of the conclusion itself. More generally, when logical omniscience holds locally, the condition is equivalent to knowledge of the material implication from the conjunction of the premises to the conclusion.

One of several respects in which these examples are untypical of epistemic logic is that very little of my work in the field uses the feature that from a purely technical point of view is what distinguishes epistemic logic from ordinary modal logic, namely a variety of indices on the knowledge operators, which are interpreted as distinguishing the knowledge of different agents, or perhaps the knowledge of one agent at different times, or both. I focus mainly on obstacles to knowledge that arise even for a single isolated agent at a single time and are best studied in that simple setting, at least in the first instance. By contrast, the majority of work in epistemic logic over the past twenty years has concentrated on obstacles to knowledge that essentially depend on variations in epistemic perspective: for instance, I know that I know something and you know that you know it, but I don't know that you know it. In order to isolate and study such interpersonal or diachronic obstacles, it sometimes makes methodological sense to idealize away most intrapersonal synchronic obstacles, as is done by the standard assumption that the epistemic logic for any one agent at one time is S5, which validates both positive introspection and negative introspection (if one does not know something then one knows that one does not know it). Such work is often admirable. But it would be a silly mistake to be induced by its concentration on interpersonal or diachronic obstacles to deny the real existence of the intrapersonal synchronic obstacles that make S5 such a huge idealization about the knowledge of one agent at one time.

3. What is the proper role of epistemic logic in relation to other disciplines, for instance mainstream epistemology, game theory, computer

sciences or linguistics?

I will discuss the case of mainstream epistemology, since it is the one I know most about. I leave it to others to decide how much of what I say would apply to the other cases too.

One's view of the proper role of epistemic logic in relation to mainstream epistemology will depend on whether one regards principles such as logical omniscience and negative introspection as true of the knowledge of real agents. If they are, then a strong epistemic logic such as S5 is the right background logic for epistemology, and it will have a powerful role to play in advancing and disciplining epistemological discussion. Like most philosophers who have considered the matter, I regard both logical omniscience and negative introspection as obviously false of real agents. Logical omniscience is false because the result of proving a new theorem is that we know more truths of logic or mathematics than we did before. Negative introspection is false because we often make errors and falsely believe ourselves to know things that are in fact false: although we don't know them (because what is known is true), we don't know that we don't them (because it seems to us that we do know them). These elementary points are as old as epistemic logic, or older; it is disturbing how often one still meets epistemic logicians who seem unable to grasp them properly. Replies to these objections tend to be leaps from the frying pan into the fire, such as the bizarre claim that mathematicians don't know what their theorems mean before proving them (in the case of logical omniscience) or disastrous concessions to old-fashioned scepticism about the external world (in the case of negative introspection). The partition model of an agent's knowledge at a time (which is what validates S5) seems to have a strangely seductive appeal to some epistemic logicians who lack the epistemological training to recognize the problematic nature of the assumptions underneath.

Many epistemologists go to the opposite extreme, dismissing epistemic logic as based on blatant errors. They neglect some alternative options. One is to redefine the subject matter of epistemic logic so that it no longer deals with the real knowledge of real agents. That may be appropriate for some applications, but it is not very promising for applications to epistemology, since it tends to obscure or lose altogether the connections with the phenomena epistemologists are trying to understand. The most obvious instance is that stipulating an interpretation that validates logical omniscience does not help us understand how the process of deduction extends knowledge. Moreover, stipulating such an interpretation is harder than is often realized. For example, merely stipulating that we are talking about ideal reasoners is insufficient, for the standard version of logical omniscience in epistemic logic implies that if some premises entail a conclusion, then *the agent knows that* if it knows the premises then it knows the conclusion: but being an ideal reasoner does not entail that one knows that one is an ideal reasoner. Usually, the more we reinterpret the formal apparatus, the more we lose touch with the epistemological applications.

A more open approach is not to reinterpret but to admit that the assumptions are literally false and to treat them like idealizations in the natural sciences. If several phenomena always occur together in practice, we may nevertheless gain insight into them by a theoretical study of each in isolation from the others, using idealized models in which the others are assumed not to occur. For instance, if one is interested in the epistemic effects of limits on our powers of discrimination, as I have been, then it makes sense to assume logical omniscience, even though it is false of all real agents, since otherwise it is hard to discern the specific effects of limited discrimination. The use of the partition model of the knowledge of an agent at a time can be defended similarly for studies of interpersonal or diachronic effects. Ultimately, of course, we want to model all the phenomena and their mutual interactions together, but that may be unfeasibly complicated to do in the short term or even the long term.

Toy models also provide a useful test of the coherence of an approach. If it does not collapse into triviality or absurdity under grossly simplifying assumptions, then it is already worth taking seriously. This is particularly valuable in philosophical discussion, where proposals are typically tested by an unsystematic search for counterexamples. I have nothing against counterexamples, but it is often helpful to apply more systematic tests too. In *Knowledge and its Limits*, I tried to show how epistemic logic can serve epistemology in such ways.

The traffic is two-way, of course. Epistemic logic is primarily applied logic, in the sense that it is driven by applications rather than by purely internal problems. Thus its progress depends on a flow of challenging new applications. Its original stimulus came from applications to mainstream epistemology, but in recent years they have tended to be submerged by applications in new fields. Although there is plenty of potential for new applications of epistemic logic to epistemology, it takes a deeper knowledge of epistemology to find them than was once the case.

Formal epistemology — of which epistemic logic forms one part — is a growing field, and it has growing interaction with mainstream epistemology, but there is still a long way to go. Striking differences in academic culture between the two sides make it hard to combine sophistication in formal epistemology with sophistication in mainstream epistemology. Indeed, although the rapid growth in work on applications of epistemic logic to fields such as computer science and theoretical economics has had excellent results, it has also tended to exacerbate the difficulties of communication between mainstream epistemologists and epistemic logicians, since the latter belong to an intellectual community that is no longer dominated by philosophers and predominantly sited in philosophy departments. Of course, there are such clashes of style in many areas of intellectual life: the macho technicians look down on the humanists for their inability to produce technical results, while the voluble humanists look down on the technicians for their inability to justify their technical assumptions. A subtler but related difference is that technical logicians are less used than philosophers to living with profound disagreements between professionals over questions within their field. In epistemic logic this unease with disagreement is sometimes manifested in attempts to resolve epistemological disputes by providing each side with a class of models that validate its claims; since the technical facts are not in dispute, this is supposed to show that each side is correct about its intended domain. From an epistemologist's point of view, such attempted resolutions look quite superficial: consistency does not imply truth, and the focus of the dispute is just shifted unhelpfully to the relation between the formal models and the intended interpretation. After all, one cannot resolve the disagreement between evolutionists and creationists by providing mathematically well-defined models of evolution and creation.

There is no reason to expect such difficulties of communication between epistemic logic and mainstream epistemology to dissolve. But they are not insuperable. Although not very many people combine sufficient expertise on the two sides, enough do for progress to be made through the interaction of the two.

4. Which topics and/or contributions should have had more attention in late 20th century epistemic logic?

Again, I will concentrate on epistemic logic in respect to its relevance to mainstream epistemology, without prejudice to the importance of topics relevant to other applications. I expect other contributors to come up with radically different lists, both from mine and from each other's. Naturally, the topics I mention are not ones to which *no* attention was devoted in late twentieth century epistemic logic; they are simply ones to which *not enough* attention was devoted.

As should be clear from my answers to the previous questions, many of the topics most relevant to mainstream epistemology deserved more attention in late twentieth century epistemic logic. The most salient of these are various ways in which S5 is inadequate as an epistemic logic for a single agent at a single time, such as failures of logical omniscience through limited powers of computation and failures of positive and negative introspection through limited powers of discrimination. As I have emphasized, it is sometimes legitimate to idealize such failures away, but we also need to study applications for which those idealizations are not legitimate. But even in major areas of mainstream epistemology which explicitly and centrally involve many agents or many times, epistemic logic has had little impact in recent years. For instance, the epistemology of memory concerns the relation between an agent's knowledge at different times, the epistemology of testimony concerns the relation between the knowledge of different agents, and the epistemology of social knowledge concerns the relation between what is collectively known and what is individually known in a community of agents. Epistemic logic may have missed some tricks through not getting more involved in those areas.

Late twentieth century epistemic logic has also devoted insufficient attention to the interaction of epistemic operators with non-epistemic operators (other than truth-functional ones, of course).

The most striking case is quantified epistemic logic. Quantifying into the scope of epistemic operators raises many difficulties, but it seems to make sense; 'How many people do you know to have a key to this door?' can be a meaningful question. The most urgent need is less for further technical developments than for more discussion of the interpretative options and their consequences for epistemic logic in the light of current approaches to the ascription of propositional attitudes.

Another case is epistemic logic extended by non-epistemically interpreted modal operators. The Church-Fitch argument mentioned above already raises subtle issues about the interaction of such pairs of operators, concerning the relation between shifts in the world at which the knowledge ascription is evaluated and shifts in the world at which the content of the knowledge is evaluated. But 'hard cases make bad law': methodologically it would be much sounder to assess such tricky examples on the basis of a general theory of modal epistemic logic, developed to account for a wide range of clearer data.

A third case is temporal epistemic logic, epistemic logic extended by explicit temporal devices, either tense operators (so that it raises some of the same issues as modal epistemic logic) or quantifiable subscripts for time on the epistemic operators (so that it raises some of the same issues as quantified epistemic logic). Philosophically, I prefer such an approach over a 'dynamic' semantics for epistemic logic itself, since it effects a more perspicuous and explicit separation of what are in principle the independent dimensions of time and knowledge. One example: temporal epistemic logic provides a natural setting for discussion of the semi-automatic updating of knowledge with the passage of time — an application of significance for both philosophers and non-philosophers.

The interest of these combinations is primarily philosophical rather than technical. From a technical point of view, there is by now much more interest in developing a general theory of combinations of logics than in studying one particular combination. But from the point of view of mainstream epistemology, it is the details of the particular combinations that matter.

5. What are the most important open problems in epistemic logic and what are the prospects for progress?

In epistemic logic, as in most branches of logic once they reach maturity, an increasingly wide divergence emerges between the problems of most technical interest and the problems of most philosophical interest. My guess is that as epistemic logic continues to develop, it will do so mainly in response to the problems of most technical interest rather than the problems of most philosophical interest. That may be for the best, even though I wonder whether any technically well-defined open problem in epistemic logic is of great importance to mainstream epistemology. From the perspective of the latter, the most important open problems are also some of the vaguest, since they concern the most fruitful ways of formally modelling phenomena of epistemological interest, such as the extension of knowledge through deduction. Some of these problems have been around for many decades, which does not give one much confidence in the prospects for progress.

As a curiosity, I will mention four technical problems that I left open in earlier work on epistemic logic without regarding them as hopelessly difficult. I won't claim them to be especially important, but maybe someone can have some fun solving them. One question is whether the system of intuitionistic modal epistemic logic that I constructed in 'On Intuitionistic Modal Epistemic Logic' (*Journal of Philosophical Logic*, 1992) is decidable. Another is whether the condition on (epistemically interpreted) modal logics I showed to be necessary for the computational constraint called 'r.e. quasi-conservativeness' in Theorem 3 of 'Some Computational Constraints in Epistemic Logic' (in S. Rahman et al., eds., *Logic, Epistemology, and the Unity of Science*, 2004) is also sufficient. A third is how far the results on iterated probabilities in epistemic logic proved for finite frames in Appendix 4 of *Knowledge and its Limits* extend to infinite frames. A fourth is to determine the logics that result from closing the modal logics KB and KTB under the double cancellation rule discussed in 'Some Admissible Rules in Modal Systems with the Brouwerian Axiom' (*Journal of the Interest Group in Pure and Applied Logics*, 1996), a rule that formalizes the claim that verification-conditions and falsification-conditions jointly determine truth-conditions. If anyone has made further progress on any of these problems they haven't told me about it, and I would like to know the answers!

Here is a much more general but still epistemologically motivated problem on which progress can and should be made. As I have emphasized, results in epistemic logic tend to be stated for setups in which the epistemic logic of a single agent at a single time is assumed to be S5, even though this involves massive idealization along several dimensions. If we drop the assumption of negative (or even positive) introspection, but retain that of logical omniscience, the result is a class of fairly tractable modal logics, the normal ones (with the T axiom that what is known is true). For a given result in epistemic logic, originally proved under the restrictive assumption of S5, we can therefore ask: how far can we generalize the result by weakening the assumption of S5 to a broader class of normal modal logics? Even for those who are much more sympathetic to the assumption of S5 than I am, it should still be a matter of interest to determine how much of it is needed for the result of interest. Which results are highly sen-

sitive to the assumption of S5, and which are more robust? The
answers should be of both technical and philosophical interest. In
this way, we can turn the negative points against S5 in a more
positive direction.

A less formal consideration along related lines is this. Work in
formal epistemology, including epistemic logic, has tended to take
for granted an epistemological outlook that is Cartesian or 'internalist' in spirit. It assumes that what one knows best is one's
own present mental states. Positive and negative introspection are
amongst the most extreme manifestations of this outlook. The
partition models that validate S5 are implicitly motivated by a
picture on which one is always in a position to know what is part
of one's evidence, and what is not. Yet such assumptions look
quaintly old-fashioned from the far more externalist perspective
of much contemporary epistemology and cognitive science. The
extreme simplicity of partition models made them a good place
to start, but epistemic logic has developed far enough for reliance
on them to be no longer technically necessary. We need to rethink
formal epistemology in general, and epistemic logic in particular,
in order to identify internalist assumptions wherever they are being made, and to rework definitions and results in order to free
them from such restrictions. I did some work along those lines in
Knowledge and its Limits. The point is not that we should replace
the internalist assumptions by externalist ones. It is enough for
the logical framework to be neutral on the issue. But epistemic
logicians need to understand what the issue is, so that they can
learn to recognize when they are making internalist assumptions,
or externalist ones for that matter. Until that is achieved, epistemic logic will remain epistemologically naïve.

About the Editors

Vincent F. Hendricks is Professor of Formal Philosophy at the University of Copenhagen, Denmark and Columbia University, NYC, USA. He is the author of many books and papers on formal epistemology, logic and methodology. He is editor-in-chief of *Synthese* and *Synthese Library*.

Olivier Roy is postdoctoral researcher at the Faculty of Philosophy of the University of Groningen (The Netherlands). He completed is PhD in February 2008 at the Institute for Logic, Language and Computation in Amsterdam, under the supervision of Johan van Benthem (Amsterdam and Stanford) and Martin van Hees (Groningen).

www.ingramcontent.com/pod-product-compliance
Lightning Source LLC
Chambersburg PA
CBHW021805220426
43662CB00006B/189